SYMBIOGENESIS

Young Boris Kozo-Polyansky, ca. 1915. (Courtesy Viktor Yakovlev, Voronezh, Russia, grandson of B. M. Kozo-Polyansky)

SYMBIOGENESIS

A New Principle of Evolution

BORIS MIKHAYLOVICH KOZO-POLYANSKY

1924

Translated from the Russian by Victor Fet

Edited by Victor Fet and Lynn Margulis

HARVARD UNIVERSITY PRESS

Cambridge, Massachusetts, and London, England

2010

Library of Congress Cataloging-in-Publication Data

Kozo-Polianskii, Boris Mikhailovich, 1890–1957.
    [Novyi printsip biologii—ocherk teorii simbiogeneza. English]
    Symbiogenesis : a new principle of evolution / Boris Mikhaylovich Kozo-Polyansky ;
translated from Russian, Victor Fet ; Victor Fet and Lynn Margulis, editors.
        p. cm.
    Includes bibliographical references and index.
    ISBN 978-0-674-05045-7 (alk. paper)
    1. Symbiogenesis.   I. Fet, Victor.   II. Margulis, Lynn, 1938–   III. Title.
    QH378.K6913 2010
    576.8'5—dc22   2009041101

This belated English translation of the young B. M. Kozo-Polyansky's magnum opus is dedicated to the memory of the great Armenian-Russian botanist Armen Takhtajan (1908–2009), without whose prodigious knowledge and enthusiasm it would not exist

Contents

The Importance of B. M. Kozo-Polyansky's Work for Modern Science

The acknowledgment of Darwin's principle of natural selection as one cause of adaptive evolution is especially important in the history of evolutionary thought. However, the explanation of evolutionary phenomena requires additional principles, one of which is Kozo-Polyansky's symbiogenesis.

The concept of symbiogenesis is one of the most fundamental explanatory factors of evolutionary change. Kozo-Polyansky recognized that complexity in individuals occurs not only through differentiation in single beings but also through the union of organisms; in symbioses, interdependence increases. The symbiosis process may even lead to the emergence of a complex life-form that displays all morphological and physiological qualities of its components in a new entity. Kozo-Polyansky, using the term coined by Merezhkovsky (1909), named this evolutionary principle "symbiogenesis."

Boris Mikhaylovich Kozo-Polyansky (20 January 1890–21 April 1957), an outstanding evolutionary biologist and a prominent specialist in the morphology and phylogeny of plants, published a number of pioneering works. His expertise led him to new general theoretical works on various aspects of evolution. In spite of its breadth, Kozo-Polyansky's work (Kozo-Polyansky, 1923, 1925, etc.) was always directed toward the expansion of classical Darwinism.[*]

*See K. M. Zavadsky, *Razvitie evolutsionnoj teorii posle Darvina (1859–1920-e gody)* [The Development of Evolutionary Theory after Darwin, 1859 to the 1920s] (Leningrad: Nauka, 1973); A. B. Georgievsky and L. N. Khakhina, *Razvitie evolutsionnoj teorii v Rossii* [The Development of Evolutionary Theory in Russia] (St. Petersburg: Feniks, 1996).

Kozo-Polyansky first stated his symbiogenesis idea at the All-Russian Congress of Botanists (1921) in the context of plant phylogeny and systematics. In his book, *New Principle of Biology: The Essay on the Theory of Symbiogenesis* (1924), Kozo-Polyansky presented a complete picture of his views on the role of symbiogenesis in evolution.

This principle had three major components. First, he defined and outlined the basic ideas. Second, he defended these ideas with information extracted from published literature. Third, he connected symbiogenesis to Darwin's principle of natural selection. Here we see Kozo-Polyansky's innovative spirit and the unmistakable passion for novelty that characterizes all of his work.

Kozo-Polyansky defined symbiosis as an important strategy for the survival of living organisms; he discussed the degree of integration of the symbiotic components. He saw that through integration subjected to natural selection, adaptive value could dramatically increase. He suggested that the highest degree of integration of symbionts leads to the emergence of united systems (consortia). Thus, the reality of symbiogenesis was embodied in the phylogenetic history of consortia.

Kozo-Polyansky, who knew several European languages and was well versed in many areas of biology, collected an overwhelming amount of information about consortia, symbio-organs, symbio-tissues, and other traits of symbiotic origin in many plants and animals. He marshalled facts to "inevitably and decidedly" witness the role of symbiogenesis in evolution.

Symbiogenesis as a factor in evolution is a "specific new principle of biology" rather than a "general biological law." It was the combination of Kozo-Polyansky's symbiogenesis and Darwin's natural selection that explained many evolutionary phenomena; a primary one is the origin of eukaryotes, a major branch of life.

Thus, B. M. Kozo-Polyansky refuted the view that symbiogenesis is a "fantasy," a "nonscientific" concept. When he placed it among well-documented scientific theories, he predetermined its significance for modern science.

This publication of Kozo-Polyansky's book in English, the best possible monument to its author, signals recognition of the importance of his symbiogenesis principle.

Dr. Liya N. Khakhina, historian of evolutionary science in Russia
St. Petersburg Branch of the S. I. Vavilov Institute of History
of Natural Science and Technology
St. Petersburg, Russia

Introduction

It is difficult for us to imagine today that more than eighty-five years ago—before we knew much about the structure of cells, before electron microscopy and many other modern tools to which we are accustomed were developed, before the principles of molecular biology, and long before the fundamental differences between bacterial (prokaryotic) and nucleated (eukaryotic) cells were explicitly described—the Russian botanist Boris Kozo-Polyansky brilliantly outlined the symbiotic origin of cells with nuclei.

At a time when it was barely possible to visualize any details of chloroplasts and mitochondria—indeed they were even confused with each other and both were often called "plastids" (Greek for "buds")—Kozo-Polyansky recognized these classes of cell organelles as former bacteria and claimed they were "symbiotic microorganisms." This young man, who lived in the provincial town of Voronezh, an agricultural center in what is known as "black earth" Russia, viewed the complex cells of plants and animals (now, of course, called eukaryotes) as consortia of cohabiting parts that had begun from very different origins as free-living bacteria ("bioblasts") that became microbial symbionts.

Only a half century after he wrote this astounding book did Kozo-Polyansky's theory come back into fashion. Only after the experimental approaches that he lacked were applied to test hypotheses that had originally been his, unknown at the time to anyone in Europe or the Americas, was his "symbiogenesis–natural selection" idea properly and independently expounded by scientists in the professional literature. After decades of neglect,

ridicule, and intellectual abuse, his ideas are now held to be correct by virtu-
ally all biologists.

In crafting his "theory of symbiogenesis," Kozo-Polyansky demonstrated a
thorough knowledge of the literature in several scholarly languages. His sound
classical education included an understanding both of the philosophical un-
derpinnings of science and of his Russian predecessors: Andrey Sergeevich
Famintsyn, who attempted to grow isolated chloroplasts in the laboratory, and
Konstantin Sergeevich Merezhkovsky (who invented the term "symbiogenesis" and wrote extensively in several languages on its power to generate novelty
as still exemplified in modern symbioses). Although the early investigators on
whose published work Kozo-Polyansky drew lacked modern tools and an understanding of genetic expression, many were keen observers. They enjoyed, or
maybe suffered, *"Sitzfleisch"*: the patience, curiosity, passion, and interest to sit
and watch microscopic phenomena for whatever time they took. They saw
sexuality, including genesis of genders and sexual organs, gamete formation,
fertilization, and embryonic development, as well as many examples of microbial ecological succession and modes of symbiont fusion. Collectively, Kozo-
Polyansky's colleagues and correspondents revealed patterns that he brilliantly
wove into a coherent theory. He well understood how much debate he would
generate in its enunciation, but today we see clearly that his evolutionary view,
like that of Darwin, has stood the test of time. The accumulation of many lines
of evidence verifies the main thrust of his postulates and the accuracy of his
details.

B. M. Kozo-Polyansky correctly drew attention to the existence of many
coevolved systems that involved mergers, temporary or permanent, of vastly
different kinds of organisms. He saw the analogies between them and the
coevolved, symbiotic systems that make up eukaryotic cells.

Kozo-Polyansky, in short, a brilliant and original thinker, was a synthesizer.
It is entirely fitting that his seminal work, *The New Principle . . .* is presented
here in an outstanding translation by Russian scholar, poet, zoologist, expert
in scorpions and their mitochondria, and now professor in Huntington, West
Virginia, Victor Fet. The book is of interest to biologists and ecologists generally, and especially to anyone interested in the evolution of life and the history

of science. Certainly the lack of appreciation of Kozo-Polyansky's concept in western Europe and North America stems partly from its extreme originality. But it also comes from the inaccessibility we Anglophones have to foreign languages. In particular, we generally lack knowledge of Russian biological literature. Much credit is also due Lynn Margulis, who has devoted a lifetime to exploring the symbiotic nature of the complex cells, the nucleated or eukaryotic components of plants, animals, and fungi, and the diversity especially of the free-living eukaryotic microorganisms that comprise probably over thirty phyla of protoctists: all the diatoms and other algae, the slime molds, slime nets, and "water-molds," the foraminifera, testate amoebae, ciliates, and so forth.

The extraordinary evolutionary and ecological role of the non-nucleated tiny forms now recognized as prokaryotes (bacteria in the wide sense)—the role of the underlying professional "planetary geochemists" that display astonishing metabolic diversity and tenacity in the face of conditions we animals consider extreme—was already hinted at by B. M. Kozo-Polyansky. We know now that it is the microbial underlayer that generated our nucleated cell ancestors, as the young Russian shows in his strikingly modern work of 1924. Prior to the chart of Edouard Chatton that listed "cellules procariotíques" and "eucariotíques," before the modern Dutch school at Delft (Albert Jan Klyuver, Cornelius van Neil, and their North American students and colleagues such as Sorin Sonea, Roger Stanier, Michael Douderoff, and Edward Adelberg), Kozo-Polyansky realized that microbial life in community, not as individuals, had produced the fabric of the living world. Would he not be delighted today to recognize that he had reconstructed over two billion years of evolution prior to the origins of either animals or plants?

On the basis of Victor Fet's translation, stimulated by Professor Armen Takhtajan's annoyed impatience and Dr. Liya Nikolaevna Khakhina's book (1992), what ideas can we rightfully credit to Boris Kozo-Polyansky?

1. The term "cell" always meant "eukaryotic cell" to Kozo-Polyansky and his international colleagues in the early twentieth century. Yet he argued that life itself preceded "cells" in the form of bioblasts, cytodes (including "flagellated cytodes" capable of locomotion), and oxygenic photosynthesizers then called Cyanophyceae ("blue-green algae," now cyanobacteria). Today we recognize these photosynthetic, oxygen-producing small life-forms as "bacteria." B. M.

Kozo-Polyansky, just as we do now, saw such microscopic beings as modern descendants of the earliest life from which plants and animals evolved.

Bacteria in the broad sense are diverse in their chemical activities. Many are capable of surviving extreme heat or freezing. Some resuscitate after insult, to form rods, filaments, or spheres under conditions that impeded their growth: lack of water, oxygen, or food. Kozo-Polyansky noticed that their growth and metabolism tended to alter their immediate environment.

Such "cytodes" or "bioblasts" appeared to him to be simpler than cells with nuclei but nevertheless to have all the properties of life. Therefore, it seems to me that Kozo-Polyansky recognized and named the salient differences between the several forms of prokaryotic life and the nucleated cells of eukaryotes. By the time he wrote his book in 1923, he already understood the ecological influence of prokaryotes.

2. Cells (of course he meant nucleated cells) are composed of organelles. In some of them, the green dots originated as "cyanophyceae." When they were taken into larger organisms, that is, acquired by nonphotosynthesizers as food, some were incorporated but not digested. United, photosynthesizer victims became permanent acquisitions of those that had attempted to digest them. The outcome? The would-be eater, whether animal, fungus, protist, or other nonphotosynthesizer, acquired the spectacular property of photosynthesis. Therefore, Kozo-Polyansky recognized symbiogenesis as the mode of positive evolution of chloroplasts of plants and algae as well as the acquisition of photosynthesis by sponges, coral animals, and the fungi that become lichens. All these groups and many more he explicitly mentioned.

3. Kozo-Polyansky, unlike his predecessor symbiogeneticists, did not dismiss Sir Charles Darwin's "descent with modification" or Darwinian "natural selection." But he united the concept of symbiogenesis, a creative process, with Darwin's mode of elimination: natural selection. Thus, in principle, evolutionary change could occur quickly (from a geological perspective) over a few generations and not require millions of years. Suddenly (not gradually) such novelties as green sponges, green slugs, and green medusoids could be generated. B. M. Kozo-Polyansky saw clearly how, by symbiotic combination, significant biological novelty would be generated and retained by Nature herself because of Darwin's selection process. He agreed with his predecessors A. S. Famintsyn and K. S. Merezhkovsky that natural selection does not create innovation. Rather, it selects those to survive and procreate. But Kozo-Polyansky's grand vision includes Darwin's concept of elimination of organ-

isms unfit at the time for survival in their particular environment and the resulting failure to leave offspring.

Therefore, Kozo-Polyansky recognized three minimal processes necessary to the phenomenon of evolution: 1) That the major forms of behavior, metabolism, and inheritance originated before cells, e.g., in minimal units he called cytodes, bacteria, bioblasts, etc. (today's prokaryotes since "cells" meant only nucleated cells in his day). 2) The tendency toward exponential growth was present not only in these minimal units with many names (prokaryotes), but also prodigious growth potential is characteristic of the symbiotic complexes that generate cells, consortia, and other multiple organisms. Furthermore, 3) Kozo-Polyansky, unlike his Russian colleagues, accepted Darwin's central concept. Symbiotic complexes, as new individuals, like all forms of life, face limits to their population growth in their struggle for existence, which is, of course, C. Darwin's (and A. R. Wallace's) idea of "natural selection."

4. Kozo-Polyansky collected, tabulated, and evaluated examples of symbiosis. He recorded physical associations studied by his predecessors, many today under detailed experimental analysis. He especially noted the greater power of evolutionary innovation when the symbiotic partners differ greatly from one another: his examples include *Nostoc* cyanobacteria in the stem glands of the angiosperm *Gunnera*, luminescent organs in marine invertebrate chordates (tunicates), and the nidamental (luminescence) glands in squid.

Most remarkably, Kozo-Polyansky explicitly recognized the formal analogy between *meiotic sexuality* (where parents are clearly related and fusion as fertilization is followed by reduction division (meiosis) and *cyclical symbiosis* (where fusion [integration] is followed by disintegration into living components ready, perhaps the next season, to fuse again). Thus, he did not fail to recognize the analogy of haploid–diploid cycles with independent unassociated partners capable of reassociation at a later time. Therefore, I think that Kozo-Polyansky appropriately analyzed the relation and evolutionary significance of both sexuality and parasexuality in eukaryotes.

Now, to what do we refer above when we mention Armen Takhtajan's annoyed impatience with us? In 1975 at the International Congress of Botany in Leningrad, a well-attended panel session about the origins of chloroplasts was arranged by Dr. Takhtajan—at the time the director of the fabulous, extensive Leningrad Botanical Garden—to which I was invited. I had just written a paper arguing the multiple origins of plastids of the different colored algal lineages. I encouraged Lynn Margulis, then an assistant professor of the

Boston University Department of Biology, to attend and present her ideas of symbiogenesis and plastid evolution. Apparently unfunded, she begged and borrowed funds to attend. The final program was assembled by Dr. A. Takhtajan, the proper chairman, who became acutely aware that his Western guests (P.R. and L.M.) were profoundly ignorant of the symbiogenesis literature of the Russian botanists, especially one whom no one had even heard about, his mentor Boris Kozo-Polyansky. Dr. A. Takhtajan, at the time distracted by his important work on angiosperms—on which he was an expert—and his duties of hospitality at the meeting, was able to quickly translate only a small part of Kozo-Polyansky's major opus, this book. The sections he had translated included these points: (1) chloroplasts are of symbiotic origin from cyanophyceae; and (2) mitochondria (= chondriosomes, chondriochonts, plastids, etc.) evolved by symbiogenesis from oxygen-respiring bacteria. He also floored us when he showed us B. M. Kozo-Polyansky's written remark that it is not outside "the realm of possibility" that "centrosomes (including their centriolekinetosomes) originated from flagellated cytodes," by which B. K.-P. meant flagellated bacteria.

Dr. A. Takhtajan scolded us: "You Anglophones believe you originated everything in botany and evolutionary science." He admonished me, "Here, read this." He thrust his minimal translation at me. And Dr. Takhtajan was correct! How arrogant of us to believe, especially at the mid-twentieth century, when German science had dominated the nineteenth, that because we could read it, all important science was only in the professional literature of the English language!

Now, thirty-four years later, finally we are able to respond to his admonishment! We bring you the authentic manuscript in readable form. We Anglophones interested in evolution can savor Fet's careful annotated translation of the book so admired and appreciated by Takhtajan himself and generations of Russian biology students.

For those who choose to follow the advances in the literature since 1924 for specific cases of symbiogenesis to see current evaluations of the status of the symbiotic associations mentioned by Kozo-Polyansky, the editors have provided a set of additional references and commentaries adequate to the task. We note, with B. M. Kozo-Polyansky himself, that the associations between individuals who are members of different species range from casual to predictable periodic (cyclical) or permanent associations. Loose unions based on interactive behavior become permanent obligate associations accompanied by genetic transfer. So long ago such processes occurred in the transformation of symbiotic bacteria into obligate, permanent mitochondria that dwell today in the cells in the bodies of all of us.

Dr. Peter H. Raven, President
Missouri Botanical Garden
St. Louis, Missouri, USA

Note to the Reader

The word "cell" for Kozo-Polyansky and his generation always meant "nucleated cell," today's eukaryotic cell. Smaller forms of life had many names; they depended upon whether what was seen was fields of dots through a microscope ("bioblasts," "cytodes," some inside cells), or motile spheres or swimming dense points ("flagellated cytodes"), or growing, smelly mucus, scum, or foam (nepheloid, ferment). Since Kozo-Polyansky was not confused in cases where literal translation would hopelessly lose the reader, we have retained the terminology of his meaning rather than his old-fashioned language. Bacteria or bacilli distinguished as ellipsoid or rod-shaped life-forms Kozo-Polyansky sometimes called by these names. Furthermore, today's oxygenic "green plant-like" cyanobacteria ("Cyanophytes" = "blue-green algae"), he clearly realized, resembled other non-nucleated, cytode life-forms more closely than they did plants (or green, red brown algae), but the terms "prokaryote/eukaryote" ("procariotique" and "eucariotique" of Edouard Chatton, see Sapp 2005) had not yet been introduced.

We became astonished by Kozo-Polyansky's insights, diligence and originality the more we worked on this manuscript. We are eternally indebted to Prof. Armen Takhtajan, who disdainfully shoved copies of his own fragments of Kozo-Polyansky translation into Peter Raven's and my hands at the 1975 International Botanical Congress. "You Anglophones believe all science must be English or German," he admonished. "Kozo-Polyansky published your ideas long before you were born!" Without his erudition, hard labor, and language skill, this book would not exist. Peter Raven and I are equally indebted to the insights of Dr. L. N. Khakhina, whose historical-critical study led us to the details of this prodigious work, and to the persistence and talent of Prof. Victor Fet, as we expect you, dear reader, will be as well. For our other acknowledgments please see page xxxiii.

Lynn Margulis
Department of Geosciences
University of Massachusetts
Amherst, Massachusetts, USA
and
Eastman Professor, Balliol College
Oxford University, 2008–2009

Kozo-Polyansky's Life

Boris Mikhaylovich Kozo-Polyansky (1890–1957) graduated from Moscow University before the Russian Revolution of February 1917 and the Bolshevik coup-d'état that followed it (October 1917). In his native Voronezh, in 1918, he joined a new Soviet university cobbled together from the faculty of Yuriev (now Tartu) University, evacuated to hinterland Russia from Estonia. Among these professors, the chair of zoology was the evolutionist I. I. Schmalhauzen, six years older than Kozo-Polyansky, who later became famous (Levit, Hoßfeld, and Ollson 2006).

Kozo-Polyansky remained in provincial Voronezh—the heartland of "black earth" Russia—for forty years, until his death in 1957. He became quite a local figure, both academic and public, a prominent botanist and plant morphologist, knowledgeable about the local flora. A former iconoclast who challenged traditions in his youth, he settled for the precarious academic life of a dean, a vice-president of his university, a director of the local botanical garden—all important positions in the provincial Soviet bureaucracy. His energy was remarkable, his work always brisk and precise. His last opus was a textbook on plant systematics published posthumously (Kozo-Polyansky 1965).

It was from his provincial Voronezh in the 1920s that Kozo-Polyansky, in his early thirties and a very prolific writer, launched a series of semipopular "theoretical" books, including *A New Principle* (Kozo-Polyansky 1922a, 1922b, 1924, 1925). Some of Kozo-Polyansky's books were indeed overreaching—for example, in *Finale of Evolution* (Kozo-Polyansky 1922a) he lamented that humankind became such a factor of planetary dimensions that evolutionary processes had ceased. In another early work Kozo-Polyansky (1922b) was among the first (again, overlooked) to posit phylogenetic systematics approaches, similar to current mainstream Hennigian cladistics, that preceded Hennig's ideas by at least three decades.

Enamored of philosophy in general, Kozo-Polyansky produced a special book for high school teachers, *Dialectics in Biology* (Kozo-Polyansky 1925), in which he tried to bridge traditional mechanistic (antivitalistic) views with the official philosophy of new Marxist rulers, not yet completely emasculated by Soviet "philosophers" of biology in the 1930s–1950s (Kolchinsky 1991, 1999, 2001); even Friedrich Engels's *Dialectics of Nature*, the mandatory reference of future decades, had not yet been translated into Russian.

Kozo-Polyansky supported Darwinism. He saw it as the only evolutionary theory of his times (Kozo-Polyansky 1924). From a Darwinian, selectionist position, Kozo-Polyansky (1923) aggressively criticized the orthogenetic theory of *nomogenesis* by the famous zoologist Leo (Lev) S. Berg (1922) (see Levit and Hoßfeld 2005). Sadly, this polemical book of Kozo-Polyansky, directed against a maverick nonselectionist colleague, would easily become a dangerous piece of "evidence" against Berg later in the 1930s, when an "idealist" label was sufficient reason for arrest and labor camp. Among countless moments that mortified contemporaries, Kozo-Polyansky's critique of Leo Berg's "idealism" was mentioned in a *Pravda* editorial in 1939. Fortunately, Berg survived.

However, later, when his Soviet rulers launched a wholesale slaughter of Russian biology, Kozo-Polyansky did not join them. Neither is there evidence that Kozo-Polyansky's career suffered from Lysenko's vicious purges, although his name appears briefly in the inquisitory deliberations of the infamous auto-da-fé of 1948 (session of VASKhNIL, the All-Union Academy of Agricultural Sciences; see Huxley 1949; Medvedev 1969; Kolchinsky 1991, 1999; Levit, Hoßfeld, and Ollson 2006). A bizarre paper by Kozo-Polyansky (1951) appeared in the *Botanical Journal* (Moscow), titled "Against Idealism in Plant Morphology." At this deadly time, two years before Stalin's death, when his colleagues, frightened by the carnage, went mute or uttered "Michurinist" slogans, Kozo-Polyansky wrote about the idealism of Goethe and of German *Naturphilosphie*, and safely criticized out-of-reach German botanists who published during the Nazi regime.

In his *New Principle*, Kozo-Polyansky's style is often rough and emotional; his Russian is educated but often fast and bristling; his imagery is animated, more so than necessary even in popular books. A Darwinist, he was a Huxley rather than a Darwin type; Kozo-Polyansky as a fine systematic botanist was an iconoclast, a polemist, and a theoretician, who displayed a clear streak of romantic *Naturphilosophie*. Kozo-Polyansky provided his own translation from Latin as a lengthy Lucretius Carus epigraph to *New Principle*. He quoted Goethe and Schiller in German. In his iconoclastic 1922 book on phylogenetic systematics of plants he began with Schiller's famous words from *Wilhelm Tell:*

> *Das Alte stürzt, es ändert sich die Zeit,*
> *und neues Leben blüht aus den Ruinen.*
> (The old is crumbling down—the times are changing,
> And from the ruins blooms a fairer life.—Transl. by Theodore Martin)

Kozo-Polyansky's *New Principle* concluded with a powerful conflation of a Darwin passage and Galileo's apocryphal *Eppur si muove:* "Even today, for many, of course, the theory of symbiogenesis would seem paradoxical; moreover, improbable. But 'when it was first said that the sun stood still and world turned round, the common sense of mankind declared the doctrine false' [Darwin, *Origin of Species*]. And yet it does move!"

How does Kozo-Polyansky connect Darwin to symbiogenesis? This issue was detailed by Khakhina (1992). However, she did not address Kozo-Polyansky's main thesis connecting *symbiogenesis* to *Darwin's pangenesis,* to the famous Darwinian statement that the "living cell is a microcosm." Amazingly, it is *not only Darwin's natural selection* to which Kozo-Polyansky connects his ideas and pledges allegiance—it is also to Darwin's *"failed attempt to understand heredity."*

First, Kozo-Polyansky interprets Darwin's *pangenesis* theory, gemmules located in all cells and organs, as a direct precursor of symbiogenesis. This oddly resembles De Vries's "intracellular pangenesis," where nuclear genes ("pangenes") were in fact Darwin's gemmules but present only in the nucleus. For Kozo-Polyansky, pangenes, or Darwin's gemmules, are present everywhere and are inherited via infection or with egg cytoplasm.

Second, he sees Darwinian *evolution via selection* in the evolutionary success of symbiogenetic cells, consortia, organs, and organisms. For Kozo-Polyansky, who does not strictly distinguish between these levels, a symbiogenetic cell, organ, or organism is more advanced and more adapted, and thus survives better in its struggle for existence. Moreover, he writes about selection among various symbionts within a consortium—and discusses what we now call "evolution of symbiotic systems," which he calls exactly this.

In some places Kozo-Polyansky is "fishing." His claim that blood platelets are symbiotic (they are not) looks like a product of "protists" of Günther Enderlein (1925), who followed A. Béchamp's "microzymas" ideas. These were long before discredited by Pasteurian bacteriology (see Sapp 1994). We recognize that Kozo-Polyansky sometimes overgeneralizes. Yet most facts are on his side. Boris Kozo-Polyansky was *not* an experimentalist, as were contemporaries whose work he reviewed. His prominent symbiogenetic colleagues included Paul Portier in France, Umberto Pierantoni in Italy, and Paul Buchner in Germany. His Russian predecessors Andrey Famintsyn and Konstantin Merezhkovsky were both already dead in 1924 (Khakhina 1992; Sapp 1994; Sapp, Carrapico, and Zolotonosov 2002). In his review—the last of this kind in Russian biology—Kozo-Polyansky relied heavily on works of other researchers. He reviewed all available contemporary literature—not a small task as Russia had just emerged from a bloody civil war where Voronezh was in the war zone!

By 1921, Kozo-Polyansky had presented his views to the first All-Russian Congress of Russian Botanists in Petrograd. Courtesy of L. N. Khakhina, we obtained the rare abstract from this congress, which is reproduced and translated here for the first time.

Theory of Symbiogenesis and "Pangenesis, a Provisional Hypothesis."
My presentation will contain the following *new* statements (when compared to the "Symbiogenesis" brochure that I previously distributed to the participants of the Congress):

(1) The statement of Linnaeus and most biologists: *Natura non facit saltum* [*Nature does not make leaps*] is not correct since formation from two (or more) organisms of a third is a leap. *Natura facit saltum* [*Nature does make leaps*].

(2) Therefore, searches for intermediate forms, *missing links* [these two words given in English—*Ed*.], in many cases will be completely fruitless. Transitions are *not* possible between two (or more) components and their sum [a new life form].

(3) Representations of origin [of a new life form] as a true [branching] tree are incorrect since the origins of new organisms occur not only by divergence of the lineages but also by their convergence and fusion: two (or more) branches fuse [anastomose] and produce a summary issue (alga+fungus=lichen).

(4) Production of new forms of organisms through symbiogenesis reflects the way in which new forms are produced by the elements: two gases unite and form a liquid; two liquids form a solid residue; iron and sulfur form

Б. М. Козо-Полянский.

Теория симбиогенезиса и „пангенезис, временная гипотеза“.

В докладе, по сравнению с брошюрою "Симбиогенезис", которая распространялась автором среди участников съезда, — выдвигались следующие н о в ы е тезисы:

1) Тезис Л и н н е я и биологии: N a t u r a n o n f a c i t s a l- t u m теряет свое значенне, ибо образование из 2 и т. д. существ третьего суммарного есть скачек.—N a t u r a f a c i t s a l t u m.

2) Поэтому поиски промежуточных форм, m i s s i n g l i n k s и т. п. во многих случаях вполне безцельны. Между 2 и т. д., слагаемыми и их суммой переходы н е возможны.

3) Представление о родословном дереве, как о подобии на- стоящего дерева, ошибочны, так как родословие организмов выражается не только расхождением линий, но и схождением, и сращением их: две и т. д. ветви, срастаясь, дают суммарное продолжение (водоросль+гриб=лишай).

4) Формообразование по типу симбиогенезиса у организмов повторяет способ формообразования, свойственный природе эле- ментарной: дьа газа соединяясь, формуют жидкость, две жид- кости—твердый осадок, железа и сера—кристаллы не похожие ни на железо, ни на серу. Новое есть результат соеднчения старого.

"Natura facit saltum": the 1921 abstract by Kozo-Polyansky.

crystals resembling neither iron nor sulfur. The new [life form] is a result of [a permanent] combination of [two or more different] old [life forms].

It is clear that Kozo-Polyansky already in 1921 had the basic ideas of evolution by symbiogenesis combined with Darwin's natural selection. Kozo-Polyansky's "fantastic" ideas on eukaryotic cell origin were clearly derived from his general statement: a "true cell" is a system, assembled from heterogeneous parts. Thus, not only mitochondria and plastids, which he treats almost as equivalent to zoochlorellae and zooxanthellae, for Kozo-Polyansky are symbionts, but all other cell organelles are suspect: centrosomes, blepharoplasts, Golgi apparatus, and even nuclei. Moreover, since "cytodes" (Haeckel's terminology for bacteria and cyanobacteria) lack nuclei, they *must* be the ancestral organisms of symbiogenetic combination. Thus, even before Chatton (1925) "formally established" (without any symbiogenetic implication about their genesis) the terms "prokaryote" and "eukaryote," Kozo-Polyansky (1921a, 1921b, 1924) realized that this most fundamental discontinuity of life on Earth is due to the symbiogenetic, synthetic, Russian *matryoshka* (nested doll)-style structure of the eukaryotic cell.

Of many examples of prokaryotic and eukaryotic symbiogenesis Kozo-Polyansky described in detail in 1924 as associations, most are currently accepted as symbiogenetic systems closely studied at all levels from genomes to ecosystems—for example, lichens; cyanobacterial symbionts of water ferns *(Azolla)*, hornworts *(Anthoceros)*, and angiosperms *(Gunnera)*; mycorrhizae of many plants; endophyte fungi of grasses; green bacteria in turbellarian worms; bacterial mycetomes of leeches, lice, ants, bedbugs, aphids ("pseudovitellus"), cockroaches; the enigmatic *Nephromyces* of molgulid tunicates (now known to be an apicomplexan protoctist); luminescent bacteriomes of tunicates and squid; and so forth.

Kozo-Polyansky, in the end, was right; in fact, he was more correct than he could imagine. More than eighty-five years later, we place his work back where it belongs, not just as historical curiosity but as evidence of dynamic new insights into the sources of evolutionary novelty. The timeline is just right: *New Principle* appeared six years after Portier's *Les Symbiotes* (1918) and three years before Wallin's *Symbionticism* (1927)—all three prescient works; all three ridiculed, ignored, and forgotten by mainstream biology. By the early 1930s, progress on symbiogenesis had stalled both in eastern Europe and in the West. We attempt here to make amends.

"And yet it does move!"

This book was originally published as *Novyi printsip biologii. Ocherk teorii simbiogeneza* [*The New Principle of Biology: An Essay on the Theory of Symbiogenesis*], Leningrad–Moscow: Puchina, 147 pp.

We have maintained the meaning by changing terms and titles. We have taken liberties with the text with one goal in mind: to communicate Kozo-Polyansky's brilliant work and to show its relevance to modern biology.

TERMINOLOGY. The retention of Haeckel's important but confusing terms "cytode" and "bioblast" as used by Kozo-Polyansky is easily explained. Kozo-Polyansky, as did his contemporaries, *limited the definition of* cell *to what we now call "eukaryotic cell."* Cells originated by symbiogenesis in his new system. Thus, his "cytodes" are "non-cellular," prokaryotic organisms, and "bioblasts" are what we call organelles or prokaryotic cells. Our "multicellular prokaryotes" in Kozo-Polyansky's terminology are "multicellular cytodes." They are composed of multiple bioblasts. Only what we recognize as eukaryotes are, in Kozo-Polyansky's language, true uni- or multicellular organisms.

Kozo-Polyansky, of course, used the traditional animal and plant kingdoms division. This created some problems in the translation where symbioses of various protoctists (or members of the kingdom Protoctista) are described. For Kozo-Polyansky and his contemporaries, algae are plants and visibly locomotory beings such as amoebae are "animals." "Plants" include algae and fungi. For particular examples, we have clarified the taxa names in square brackets, or, if necessary, in the notes. We append the full list of taxa mentioned in the book, following the modern standard of Margulis and Chapman (2010).

We replaced Kozo-Polyansky's archaic language with more familiar terminology, much of which was already in use by the 1920s. Kozo-Polyansky rarely uses "cytoplasm" (Russ. *tsitoplazma*). Sometimes he mentions "protoplazma" but usually abbreviates it to "plazma" ("plasm"). We rendered "plazma" everywhere as "cytoplasm" but retained "protoplasm" where he used the full word "protoplazma." Kozo-Polyansky uses "chondriosomes," an outdated term we replaced with "mitochondria" (which he also used in the original).

Other equivalencies of Kozo-Polyansky's terms include "plasmatic" = "cytoplasmic"; "cyaneae" or "Cyanophyceae" = "blue-greens" (cyanobacteria); "chlorobacteria" = "green photosynthetic bacteria"; "chlorophyll granules" = "chlorophyll organelles"; "cell's center" = "centrosome" or "microtubule organizing center"; "organoid" = "organelle"; "assimilation" = "photosynthesis" or "absorption"; "sago palms" = "cycads"; "micron" = "micrometer (μm)," etc. For "blood" in insects we use "hemolymph." See the glossary that begins on page 181 for more detailed explanations.

REFERENCES. Pages 137–144 of the original Kozo-Polyansky book contain the bibliography. References were greatly simplified for the nonacademic reader. As the author says, "This list, which is parallel to the text, of course, is not exhaustive. It includes only works containing bibliography of certain issues, instead of listing all possible references." Since non-Russian authors were listed in text only in Russian transliteration, it was often difficult to confirm the names, especially for obscure publications. Most works were referenced only by author's name, without even the year of publication. Although we tried to exactly verify the publications, discrepancies and uncertainties remain. Thus, not all authors mentioned in the text were included in the reference list. Kozo-Polyansky usually omitted article titles and page ranges; the text also has a large number of spelling errors, especially in foreign names and Latin names of taxa.

Our references include those cited in Kozo-Polyansky's list, and where possible we complete them, following modern academic standards. Our list includes other works mentioned in Kozo-Polyansky's text but not listed in his bibliography. Especially when Kozo-Polyansky referred to a general concept, his references to authors were matched to a specific work with limited certainty.

FOOTNOTES AND ENDNOTES. Kozo-Polyansky's original footnotes were retained, referenced by asterisks. The endnotes referenced by superscript numbers were added by the editors and are given in the Editors' Commentary.

An additional list of modern references was compiled by the editors. Even a cursory review of the enormously diverse field of current symbiosis and symbiogenesis research is impossible in a work of this size. By providing brief references to the modern symbiogenetic science literature, we tried to place Kozo-Polyansky's examples in the context of our new century. More details are in the informative chapter on Kozo-Polyansky by Khakhina (1992). The goal here is to direct the English-language readers to samples of the ongoing research on symbiogenesis.

The reason such new detailed information is widely available now is that Kozo-Polyansky's boldly stated symbiogenesis, his "new principle," has joined the "mainstream" of environmental and evolutionary science. Hundreds of scientists labor fruitfully on its behalf, many of whom published in the Balaban journal *Symbiosis* (editor-in-chief Professor David Richardson, St. Mary's University, Halifax, Canada) or attend meetings of the International Symbiosis Society (ISS). Scientific research since the publication of *A New Principle* has vindicated, substantiated, and expanded *most* of Kozo-Polyansky's bold statements. Where Kozo-Polyansky was incorrect he was misled by the enthusiasm of other investigators such as Portier (1918); see Editors' notes and Sapp (1994; 2009) for details of both Wallin's and Portier's mistaken claims that they grew mitochondria in culture. Kozo-Polyansky's ideas now thought to be wrong remind us that, isolated in the Russian provinces, he was groping in the dark. Relevant data accumulated by bacteriologists, phycologists, botanists, zoologists, and electron microscopists, and the ultimate advent of formidable protein and DNA molecular-biological techniques, all came after Kozo-Polyansky, his symbiogeneticist predecessors and contemporaries (Famintsyn, Merezhkovsky, Portier, Wallin).

Kozo-Polyansky often overreached in his sweeping, enthusiastic philosophical musings. He sometimes was tinted or tainted by the blunt "dialectical materialism" of his political persuasion, its orthodoxy guised as revolution. However, in the end he was correct in his insistence on symbiogenesis as a major source of heritable novelty in the Darwinian evolutionary processes, and his brilliant work deserves wide recognition.

Compendia and reviews on symbiogenesis include Margulis and Fester (1991), Margulis (1993), Saffo (1992), Geus and Höxtermann (2007). Buchner (1965) is still an indispensable treatise primarily on animal symbioses with bacteria. For a review of insect mycetomes, which occupy a prominent place among Kozo-Polyansky's examples, see Douglas (1989), Bourtzis and Miller (2003, 2006). Sapp (1994, 2002) provided a detailed account of historical development of symbiogenetic theories before and after Kozo-Polyansky's time. Khakhina (1992) did the same for Russian science. Reviews of selected groups of organisms discussed in Kozo-Polyansky's book are provided in the list of Commentary References that begins on p. 163.

ILLUSTRATIONS. The original Kozo-Polyansky book had several illustrations of symbioses reproduced mainly from old papers of Perfiliev, Buchner, and Peklo. Our Commentary is complemented by new illustrative material (photographs, phase contrast and PIC light micrographs, TEM and SEM micrographs). Please refer to figure captions in combination with editorial commentary for modern information on a number of organelles and organisms described by Kozo-Polyansky; many have become outstanding models for symbiogenesis research.

Victor Fet
Department of Biological Sciences
Marshall University
Huntington, West Virginia, USA

Acknowledgments

VF is grateful to his father, Dr. Yakov Fet (Novosibirsk, Russia), who obtained a rare original copy of Kozo-Polyansky's book, with the help of Dr. Grigory Dymshits (Institute of Cytology and Genetics, Siberian Branch of the Russian Academy of Sciences). The book, located in the institute's library in 2005, had been a personal copy of the great Russian geneticist A. S. Serebrovsky (1892–1948).

Marshall University (Graduate College Fund for Reassigned Time for Research, and the former graduate dean, Dr. Leonard Deutsch) provided release time from part of VF's teaching duties from January to May 2007 for this project, without which this translation would not yet be completed. We thank Dr. Charles Somerville, the current dean of the College of Science and the former chair of the Department of Biological Sciences, for his enthusiastic support of this ambitious project. We thank the former dean of the College of Sciences, Dr. Andrew Rogerson, and the assistant dean, Dr. Wayne Elmore, for their support. VF's travel to the Margulis lab in March 2009 to work on this book was also supported by Marshall University.

Special thanks go to B. M. Kozo-Polyansky's grandchildren, Evgeniya Yakovleva (Moscow, Russia) and Viktor Yakovlev (Voronezh, Russia), for their kind permission to publish this translation according to Russian copyright laws. Viktor Yakovlev kindly provided a rare photograph of B. M. Kozo-Polyansky for this volume. Drs. Vladimir Agafonov, Maria Mikhaylenko, Oleg Negrobov, Dmitry Shcheglov, and especially Dr. Viktor Golub (all of Voronezh, Russia), greatly helped in obtaining valuable information about Kozo-Polyansky.

We were honored by the approval afforded by Dr. Armen L. Takhtajan (St. Petersburg, Russia) (1910–2009), who was arguably the best botanist in the world and the famous successor of Kozo-Polyansky's ideas in plant systematics. Takhtajan, at the International Botanical Congress (Leningrad, 1975), was also the first to inform Western scientists about Kozo-Polyansky's underappreciated symbiogenesis theory. In 1973, Armen Leonovich wrote that "it has become perfectly obvious that the symbiotic theory of the origin of the eukaryotic cell, previously received with suspicion or rejected, has acquired many supporters in recent years" (Takhtajan 1973, 1983).

We are grateful for the attention and valuable comments from Dr. Liya N. Khakhina (St. Petersburg, Russia). Liya Nikolaevna was the first to open to the Western scientific community the intricate and rich story of Russian symbiogenetics research, and especially of the hitherto completely unknown B. M. Kozo-Polyansky (Khakhina 1992).

We thank Professor Peter H. Raven for his wonderful introduction to this translation.

We also thank Elena Takhtajan, Drs. Yuri Kitaev, Yakov Gall, Tatyana Shulkina, and George Gause for their kind help, as well as Drs. William Provine, Georgy Levit, and Michael Golubovsky for their valuable comments. We acknowledge the great help of students and colleagues in the Margulis lab in this endeavor, namely, Celeste Asikainen, Dr. Michael Chapman, Kendra Clark, Dr. Michael Dolan, Sean Faulkner, Carolina Galan, James MacAllister, Melishia Santiago, and Bruce Scofield; as well as Emily Case and Dr. Jennifer Margulis di Properzio.

We greatly thank Drs. André Fortin, Mary Rumpho, Claudio Bandi, Luciano Sacchi, Mary Tyler, and Andrew Wier, as well as Jean-Marc Gagnon (Éditions MultiMondes, Canada), Nick Hall (ARS Photo Unit, USDA), and Vyacheslav Ponomarev (Russia) who shared with us wonderful illustrations used in this book. We thank Dr. Mary Beth Saffo for her detailed comments on *Nephromyces* symbiosis.

VF thanks his wife, Galina Fet, for her kind and enthusiastic help and support during this work.

Finally, we are grateful to Harvard University Press, its reviewers, editors, and the editorial director, Michael G. Fisher, and his assistant, Anne Zarrella, for their help and support. We especially thank Melody Negron, the production editor; Lisa Williams, the copyeditor; and Lisa Roberts and Heather Shaff Beaver, the designers, for their wonderful work.

All of these people contributed their attention, time, and effort to bringing Kozo-Polyansky's forgotten work to the world, more than eighty-six years after its appearance in Russia, at the dawn of modern biology.

Victor Fet

Lynn Margulis

SYMBIOGENESIS

Проф. Б. КОЗО-ПОЛЯНСКИЙ

НОВЫЙ ПРИНЦИП БИОЛОГИИ

ОЧЕРК ТЕОРИИ СИМБИОГЕНЕЗА

Издательство „ПУЧИНА"

1924

The original Russian book title page (1924).

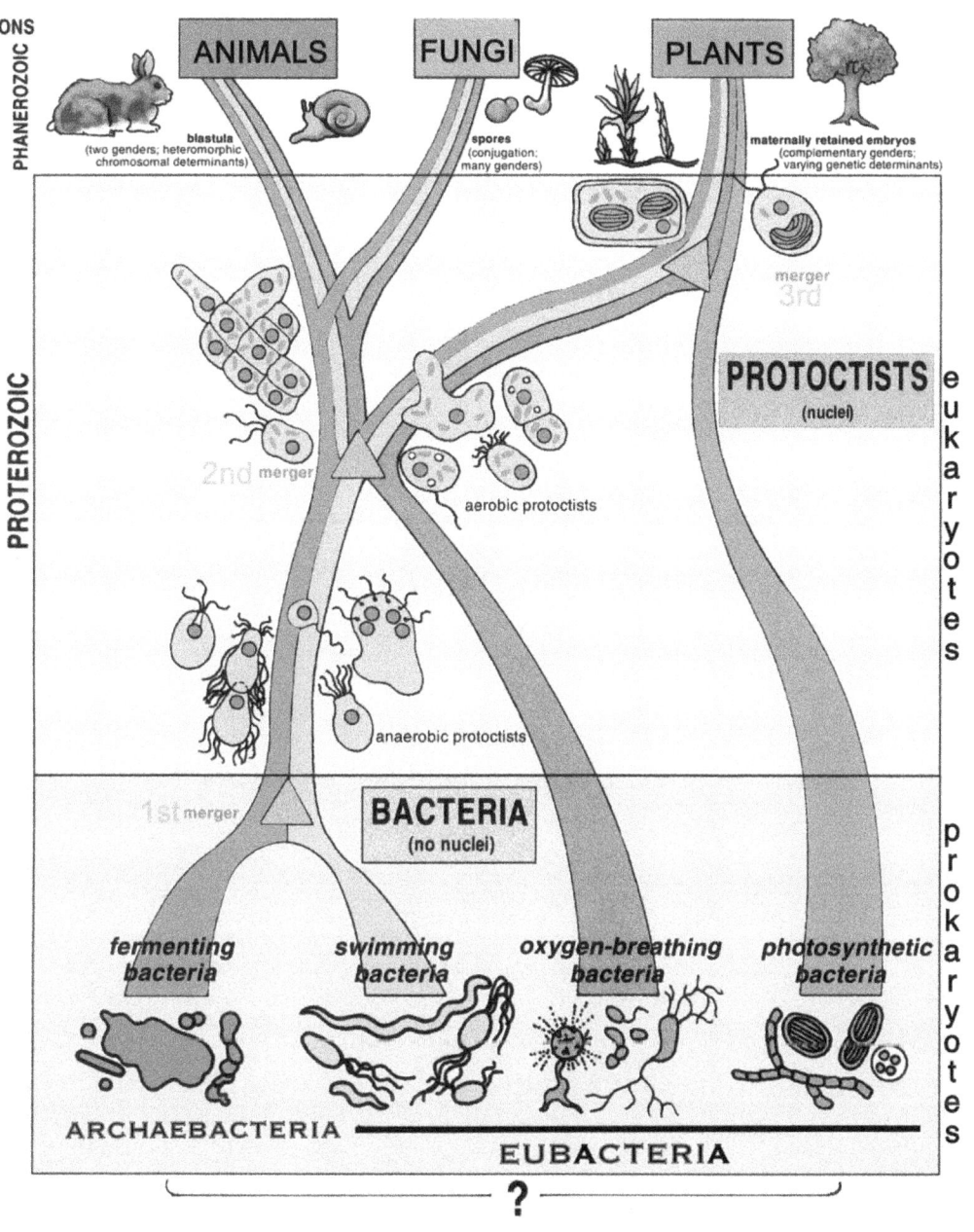

Modern status of Kozo-Polyansky's view of the net of life: Symbiogenesis refined by Darwin–Wallace "natural selection." Drawing by Kathryn Delisle.

Preface

tibi vehementer nova res molitur ad auris
accedere et nova se species ostendere rerum.
Sed neque tam facilis res ulla est, quin ea primum
difficilis magis ad credendum constet, itemque
nil adeo magnum neque tam mirabile quicquam,
quod non paulatim minuant mirarier omnes . . .

desine qua propter novitate exterritus ipsa
expuere ex animo rationem, sed magis acri
iudicio perpende, et si tibi vera videntur,
dede manus, aut, si falsum est, accingere contra.

For a new doctrine presses earnestly to approach your ear, and a new scene of things to display itself. But neither is any thing so easy, or credible, as that it may not seem rather difficult of belief at first; nor, likewise, is there any thing so great, or any thing so admirable at first, at which all men alike do not by degrees less and less wonder. . . . Wherefore forbear, through being alarmed at mere novelty, to reject any argument or opinion from your mind, but rather weigh it with severe judgment, and, if it seems to you to be just, yield your assent to it; or, if it be false, gird up your loins to oppose it.

Lucretius, De Rerum Natura (Book II: 1024–1029, 1040–1043), "literally translated into English prose" by John Selby Watson (*On The Nature of Things*. A philosophical poem. London: Henry G. Bohn, 1851, p. 92).

[The original has this epigraph both in Latin and in Russian, translated into prose by Kozo-Polyansky himself—*Eds.*]

One can think that natural sciences have now entered a zone of great reforms.

Majestic events have already unraveled in chemistry and physics, and all thinking humankind listens to their rumble. The principle of relativity and the disintegration of the atom are on everybody's lips.

However, it is not widely known that a wave of new principles also engulfs biology.

In this science as well a new and powerful creative principle has appeared that beckons it forward. Biology also undergoes a "disintegration of elements," and the idea of ancient atomists finds its unexpected development.[1]

This new theory, which may be destined to renew biology, we call "the theory of symbiogenesis." Like many, if not all, great doctrines, it was conceived long ago; it was mentioned by the great Darwin himself. However, only remarkable discoveries of the recent decade placed it in the forefront of the attention of science globally.

Every organism, beginning with unicellular organisms, represents a system of more elementary heterogeneous organisms. This inductively derived concept lies at the foundation of the new doctrine.

Since the theory of natural selection was proposed, no other idea in biology has been more universal and original, more destructive and creative than this *new principle of biology.*

Every great scientific discovery passes through three historical phases. First, the public says: this is just ridiculous. Second, they begin to say: this is contrary to established opinion. But eventually they say: we have long known it!

This is the conclusion of the famous Whewell [1840],[2] the author of the classic *History of the Inductive Sciences.*

And our doctrine, of course, is also bound to follow this road of drudgery. It has just moved to its second step.

Let this theory of symbiogenesis still be very far from general acceptance, let peaceful struggle still surround it. But everybody with some interest in life sciences or anyone interested in new scientific ideas, I think, deserves to know what the new theory is, the discoveries and observations on which it is based, and what consequences the theory of symbiogenesis has for life sciences.

To provide a brief and, as far as possible, simple answer to these questions is the goal of this book.

For their interest in my work and help with some references, I am glad to acknowledge N. I. Vavilov (Leningrad), M. I. Golenkin (Moscow), L. I. Kursanov (Moscow), V. N. Lyubimenko (Leningrad), G. Martin ([New] Brunswick), Donald Reddick (Ithaca), and J. Schramm (Washington [University, St. Louis]).[3]

I

Noncellular Organisms (Cytodes) and Bioblasts [Prokaryotes]

The *minimum* of life's organization is usually thought to be the cell. Every living being consists of cells "as a building consists of bricks" (Weiss) or, at the very least, of a single cell, most agree.

However, there are organisms called noncellular organisms, or "cytodes" (Haeckel, Heidenhein). These, for example, include Schizophyta; i.e., the bacteria and Cyanophyceae or cyanophyta, blue-green "algae" [later "blue-greens"], the simplest of all known organisms, neither yet animals nor plants [Conn and Conn 1923, Enderlein 1921, Jordan 1922, Lieske 1922, Wettstein 1921, 1923].

Some place Schizophyta among unicellular organisms. However, their bodies lack the structure of a complex device we call a cell. They lack the nucleus, always present in the cells of both animals and plants. They don't have the green chloroplasts so typical of plant cells. They lack centrosomes typical of animal cells, and other such "organs" [organelles], the presence of which is inalienably linked to the concept of "cell."

The cytode "cell," therefore, more appropriately should be called by a different name, e.g., "bioblast."[1]

1. Bacterial Bioblasts

Bacteria include the smallest living organisms. Some are so small that their presence, even in large numbers, can be determined only indirectly, e.g., by results of their life activity. It is impossible to observe them even at the highest magnification. Some bacteria are seen not as individuals but as "nebulae." Yet giant bacteria may be up to 100 μm long [Nadson and Visloukh 1923].

Bioblast shapes include spheres (cocci), elongated ellipsoids (bacteria), straight rods (bacilli), commas (vibrios), short coils (spirilla), and long coils (spirochetes).

Their surfaces can be smooth and naked, or covered by flagella that vary in number and position. Small bioblasts appear to lack internal organization.

In larger bacteria, cell walls, which are sometimes difficult to see, mainly nitrogen-containing, can be distinguished. During plasmolysis, walls either do or do not detach from the bioblast contents. The wall in *Bacterium [Acetobacter] xylinum* is made of cellulose; in other cases walls probably consist of pectin compounds.

Inside the bioblast, often after special treatment, so-called Babes bodies,[2] or metachromatic granules, present in varying numbers, are detected. These may consist of chemicals that resemble the chromosomes of cellular organisms.

Most commonly, when two corpuscles, or two groups of corpuscles, are present, they are located at the opposite poles of the bioblast. When their number is large, they may be positioned variably, without obvious order. A kind of filamentous connection sometimes can be seen between separate corpuscles.

In the same culture of a given type of bacteria, the number of corpuscles varies. Some bioblasts completely lack corpuscles. Possibly they are nothing more than products of metabolism (Alfred Fischer [1897]), or just more intensively staining areas of the cytoplasm.

The polar position of these granules probably precedes the bioblast division; earlier they are located at the bioblast equator.

In addition to metachromatic granules, Bütschli discovered alveolar structures in some large bioblasts. Of the alveoli, one that lies in the center is the largest. This type of structure is not discerned in most bacteria; alveoli can be considered, by analogy with similar formations in real cells, as "granules," i.e., dense cytoplasmic bodies, but not vacuoles, as they are often interpreted. Their granular nature is confirmed by the fact that in many cases they do not disappear when a microscopic specimen is dried, or when air is pumped out.

Along with colorless bacteria are colored or "chromogenic" ones. In some chromophore bacteria of Beijerinck [1890], the pigment is a component of the bioblast. In others the pigment body is waste product, which may easily move away into the environment or may remain for a long time near the bioblast (parachromophore bacteria). The bioblast itself may be completely colorless; in chromophore the pigment may penetrate the entire bioblast or only lie next to the surface.

The "purple" bacteria, for example, produce red "bacteriopurpurine" (which belongs to lipochromes) as well as green "bacteriochlorine." *Bacterium polychromum* contains a yellow pigment, which is not water-soluble, and a water-soluble red-purple pigment. Other yellow, blue, fluorescent, etc. pigments may be present.

Finally, considerable attention has been attracted recently by a special group of bacteria, green bacteria that possess a pigment extremely similar to plant chlorophyll. It is probably just an insignificant variation of chlorophyll, which is confirmed by spectral analysis. This pigment, for example in *Pelodictyon,*[3] according to Perfiliev [1914a], gives a "diffuse" coloration to a microbe's bioblast.

The concept that the bacterial body (bioblast) is a plant cell was formulated at a time when structural details were unknown. Later, authors who insisted that bioblasts of schizophytes were real cells fervently disagreed among themselves.

Zettnow and others considered the whole body of a bacterium as a nucleus without cytoplasm, for example, while Fischer, Migula, and others maintained that bioblasts are cytoplasm without nuclei. The metachromatic granules (Babes corpuscles) are considered concentrations of stored chemicals by Fischer, Günther, and others; "chromioles" by Ruzicka, Merezhkovsky, and others; chromosomes by Kirchensteins [1922], Swellengrebel, and others; entire nuclei by Sjobring, Benecke, A. Meyer, and others; embryonic nuclei by Bütschli; a diffuse nucleus by Gottschlich, Tischler, and others; and so on.

This controversy allows us to conclude that there is no basis for any objective analogies of a bacterial bioblast with a real cell. Bacteria lack true individual nuclei as well as other "organs" [organelles] typical of a cell.

Bacterial reproduction occurs by binary fission, i.e., parallel to their width, and is considered highly characteristic for the entire group. In some cases, pairwise fusion of bacterial bioblasts has been observed, which can be seen as an elementary instance of sexual process. Formation of internal spores (endospores) takes place, in addition.

The [bacterial] way of life is highly diverse. Most are parasites and saprophytes; some colorless forms are able to consume atmospheric carbon chemosynthetically, i.e., by use of chemical energy. Some green bacteria perform photosynthesis as do green plants.

In other green bacteria, their pigment serves to use solar energy while available organic compounds are consumed. Something similar, according to the data of Molisch [1918], is suggested for the purple bacteria.

Many bacteria are able to live in environments that lack oxygen (anaerobes); many are tolerant to high temperatures, to alkaline environments, and to various strong poisons, and can even utilize them.

A very characteristic feature of bacteria is their diverse and energetic fermentation activity—their ability to produce catalysts of a colloid nature, which speed up various chemical reactions.

2. Cyanophyceae, or Blue-Green Algal Bioblasts

Bioblasts of blue-greens,[4] due to their large size, are usually more accessible than others for study. Their shape is usually spherical or ellipsoidal. Their walls, in cases when they can be studied, contain nitrogen, and they lack flagella (however, in *Merismopedia,* Goebel [1915–1923] noticed production of zoospores). [This is not correct; although some swim and many glide, neither bacterial flagella nor undulipodiated zoospores are known in cyanobacteria—*Eds*.]

In bioblasts, one can distinguish a peripheral cytoplasmic layer, or chromatoplasm, and a central mass, or centroplasm. Blue-green photosynthetic pigment is present in chromatoplasm, diffusely distributed and penetrating through it. Centroplasm, on the contrary, is colorless. In other aspects both layers are identical. Both layers exhibit various kinds of granularity, with gradual transition between its varieties.

In the central cytoplasm of each of the bioblasts, one can distinguish either liquid or more dense formations called "endoplasts." Their number and size are not constant. Between them lie spherical, strongly refractory bodies of various sizes and in various amounts—so-called epiplasts. They also may be absent altogether. In addition, more or less polygonal granules, easily distinguished from the previous ones, can be found. They usually lie in the periphery of the centroplasm (ectoplasts of Baumgärtel [1920]).

In the most peripheral portion of the bioblast, grains or granules are positioned so that they resemble filaments parallel to the surface (fibrils of Hieronymus).

The pigment may be blue-green, olive-green, patina-green, blue, red, yellowish-brown, brown-green, or violet. It is a mixture of true chlorophyll and more or less blue phycocyanin; the latter, unlike the former, is easily soluble in water.

A number of contradictory explanations regard the blue-green "cell," just as with the bacterial "cell." Even the most recent of those opinions are diametrically opposed.

Endoplasts correspond to alveoli containing nuclear fluid; spaces between them, to lignin; epiplasts, to chromatin granules; and ectoplasts, to nucleoli of a nondividing nucleus of a typical cell, according to Baumgärtel (1920).

Geitler (1922) seriously criticized this explanation and stated that "considering them without a preconceived notion, all these 'plasts' make an impression of the products of metabolism lodged in the cytoplasm . . . accordingly, they are strongly modified in culture, and also can look very differently in the material taken from nature." All this comparison and identification, similar to that of Baumgärtel, is generated by the desire to find, in the central body of blue-greens, components comparable to true nuclei.

"There are no reasons to call this cytoplasm nuclear only due to the fact that carbohydrate granule-plasts lodged in it sometimes exhibit a grouping that resembles true, nondividing nucleus. If nuclear functions are already present in the cytoplasm, we cannot say that there 'is a primitive nucleus'; there simply is no nucleus." Geitler was supported by Haupt (1923).

Blue-greens clearly should be considered organisms that lack nuclei. They are obviously cytodes, according to Geitler and others. An expert on blue-green cytology, Fischer [1897], also denies the existence of a nucleus [in this group]. Inclusions in the central body, which have been interpreted as chromatin granules and similar characteristics of true nuclei, he considers to be products of metabolism, including a special carbohydrate named "anabaenin" [cyanophycean starch].[5]

Fischer maintains that the bioblast chromatoplasm is a single, large chromatophore, shaped as a cylinder or an open cylinder, enclosing the central body and open at least on one side. Previously, Kohl and others interpreted chromatoplasm as a collection of small spherical chloroplasts, i.e., chlorophyll organelles.

The green layer (chromatoplasm) is identical to that of the central body; it appears to be just a cytoplasmic layer containing pigment. Physiologically, in the opinion of Geitler and Haupt, chromatoplasm could be equivalent to a single chloroplast (chlorophyll organelle) of a plant cell; however, morphologically there are no reasons to consider them identical.

"Thus [says Geitler], a blue-green 'cell' exhibits the same structure as a bacterial cell." In it, one cannot demonstrate the existence of components typical for a true cell. A blue-green "cell" is, in fact, also not a cell but a bioblast.

Blue-greens are as capable of photosynthesis as true green plants. The product, however, is not starch, but a special carbohydrate—glycogen (Fischer). A remarkable property of blue-greens is not only their ability but their preference to use available organic compounds. At the same time, they utilize solar energy.

Often, blue-green "algae" live under such conditions where no green plant (such as algae) would survive—in thick, rotting pond scum, etc., similar to bacteria. They are also very resistant to temperature changes. For example, "observations demonstrated the presence of live blue-greens in the water of hot geysers of Yellowstone Park in the United States. The temperature of the surface covered with a film of blue-greens was 80°C" (Arnoldi [1901]).

Blue-greens and bacteria, therefore, greatly resemble each other. Forms placed by some competent authors among blue-green "algae" are considered bacteria by others. E.g., *Pelodictyon* is classified with blue-greens by Szafer [1911], West [1916], and Pascher [1914], while Perfiliev [1914a; Monteverde and Perfiliev 1914] argues that it is a green bacterium. All non-green forms, which are not able to photosynthesize, were once classified as bacteria. This criterion is now discarded, since many blue-green cytodes prefer a saprophytic way of life. Also, some quite typical bacteria with a green pigment (green bacteria) have been discovered. All cytodes with blue-green pigment are currently classified as blue-greens; and all others, as bacteria.

However, it is hard to decide whether a number of forms are green or blue-green. Some blue-greens have typical pigment but otherwise bacterial features, writes Pascher [1914]. They are extremely small and appear, much like the simplest bacteria, to lack any organization. In *Dactylococcopsis* and in a blue-green alga that cohabits with *Oicomonas* (see below), the bioblast has the same shape as bacilli. In *Chroostipes*, the shape resembles that of a vibrio. In two latter forms, a phenomenon has been recorded that is extremely important in a theoretical sense—the lengthwise division. This has not been previously found in cytodes and has not yet been found in colorless bacteria. [Probably not verified.—*Eds.*]

Cyanotheca longipes[6] possesses a long flagellum. And so on.

In short, blue-greens can be considered as blue-green bacteria along with green, purple, and other divisions of the bacterial group in a broad sense.

The term "microbes" will be used here, therefore, for both bacteria and blue-greens, and the term "bacteria" will be reserved for all cytodes except the blue-green ones.

Especially close to microbes in our sense are "fungi," the yeasts (Saccharomycetes). Their "cells" are usually much larger, and characteristically they reproduce by budding. They usually form four endospores; a sort of sexual process is also known when the couples of "cells" fuse. The presence of typical cell organelles has not been shown in many yeasts. The structure of more closely studied forms resembles that of large, colorless blue-greens, according to Kunstler [1889] and others. Yeasts stand close to bacteria in the broad sense, not only in their body structure but also in their way of life, especially their fermentation activity.

On the other hand, it appears that it is hard to draw a boundary between colorless bacteria and small, also colorless flagellates [mastigotes]. The former have been artificially distinguished from the latter on the basis of their division type: binary fission in bacteria, and lengthwise division in flagellates. It seems more reasonable to place into flagellates only those forms that have a typical cell structure. In particular, green flagellates are easily distinguished from Pascher's "blue-green algae" and green bacteria by having chlorophyll organelles [chloroplasts] in their cytoplasm and, naturally, by the presence of nuclei.

3. Symbiosis among Cytodes

Cytodes are usually found in colonies, e.g., aggregations of bioblasts of the same kind. Separate individuals are cemented together by mucus or gel, which forms more or less thick capsules around each bioblast. Gelatinous masses called *zoogloeas* enclose more or less significant numbers of microbes and are extremely variable in shape and size. Zoogloea may have special shared mucous layers that enclose all members of a colony at once.

Sometimes, cytodes in a colony form clumps, but often there is a characteristic beadlike formation, or a formation shaped as "coin stacks." Such stacks often end with a long, tapering, pointed projection. Several such stacks, each with its own gelatinous capsule, may be united into a thicker rope. Additional gelatinous cover may embrace all of them. A very widespread phenomenon is observed when filamentous or stacklike colonies assume a spiral or a corkscrew shape. This shape resembles that of vibrios, spirilla, etc.

The mucus is a colloidal substance that can increase in volume greatly in all these cases. Its chemical nature is insufficiently known, and it might not always be the same.

In addition to an incredibly common formation of colonies (collections of units of the same type), microbes are also very prone to form consortia, i.e., close collections or systems of heterogeneous forms. Due to a great theoretical importance of such symbiotic unions, it is necessary to get acquainted with their variants.

A CONSORTIUM OF THREE COLORLESS CYTODES. In the process of free nitrogen fixation, a very important role belongs to zoogloea found in soil and studied by Winogradsky [1895]. It comprises two types of filamentous bacteria and one *Clostridium (C. pasterianum)*, i.e., a microbe that enlarges in the shape of a spindle during spore formation.

The ability to bind gaseous nitrogen is an exclusive feature of *Clostridium*. This organism does not tolerate oxygen and is anaerobic. Bacteria cohabiting with it protect it from oxygen, since they need oxygen and also use nitrogenous compounds from their partner.

We have to admit here that the basis for close cohabitation of heterogeneous units, the basis of consortial symbiosis, is formed by mutualism—existence of mutual benefit. This mutualism, of course, has nothing in common with altruism, with conscious and voluntary mutual assistance. It is a pure "commensalism": one partner takes up what other does not need, and vice versa. A quite viable and closely knit system emerges. A consortium is able to flourish where its partners would perish on their own.

A CONSORTIUM OF TWO GREEN CYTODES. In zoogloeas of a blue-green *Anabaena*, between the chains and nodules of its bioblasts, Pascher

[1914] discovered minuscule (about 1 μm in diameter) blue-green or green microbes, *Cyanodictyon endophyticum*, clumped in aggregates.[7]

These aggregates, at greater magnification, are weblike colonies; loops of this network consist of chains of spherical individuals.

While *Anabaena* is green in its normal condition, its smaller partner usually has a pale coloration and, probably, is not photosynthetic. Toward summer, as *Anabaena*'s filaments disintegrate, its symbiont acquires more intense green coloration. Finally, only propagules of *Anabaena* remain. By this time, *Cyanodictyon* aggregates disappear as well. Single individuals of *Cyanodictyon* are later seen attached to the surface of young *Anabaena* propagules. In this way, *Anabaena* is accompanied by *Cyanodictyon* throughout its life history.

Physiological relationships between partners in this consortium are not yet known. Most likely, *Cyanodictyon* may obtain its food from the mucus of *Anabaena;* however, the benefit for *Anabaena* is not clear. One-sided "egotistic" [selfish] exploitation may take place here. *Cyanodictyon* may use the "refuse" of its cohabitant, i.e., not be a parasite but just a "commensalist."*

A CONSORTIUM BETWEEN A COLORLESS CYTODE AND GREEN CYTODES (combination of a colorless individual and a green colony). The central position is taken by a spindle-shaped colorless bacterium ca. 3–4 μm in length, with one long flagellum [probably a bundle of flagella as individual bacterial flagella are beneath the limit of resolution of light microscope—*Ed*]. Its mucous covering includes a large number of rod-like green bacteria positioned in longitudinal rows. "Reproduction of green bacteria takes place along with growth of the central organism. When the central organism divides (by

---

*Let us agree on terminology. *Mutualism* is a symbiosis based on mutual exploitation. *Egotism* is a symbiosis based on the exploitation of one partner by the other. In both cases we can have: (1) *parasitism* (antibiosis, antagonism), when one partner exploits another forcefully, depriving the latter from items it needs for itself; and (2) *commensalism*, when one partner picks up the "refuse," or "leftovers" [waste products] remaining from the activity of the other partner (which could be not only food, but excreta, enzymes, gases, etc.).

Domatism is the exploitation of one partner by the other as a shelter, a "living space" in order to escape predators, temperature, desiccation, oxygen, etc. Domatism is essentially egotistic but the landlord could obtain other types of benefits from its tenant.

We should reiterate that *mutualism* has nothing to do with *altruism* and the so-called *harmony in nature* since altruism implies voluntary and conscious mutual help, while mutualism happens automatically and mechanically, unconsciously, and is based on the chance coincidence of mutual interests.

binary fission), its surface is covered by green 'cells' [i.e. bioblasts—B. K.-P.] distributed equally between daughter individuals."

In some cases, division of offspring consortia starts earlier than their dissociation, which results in combinations of four, or more rarely six, "individuals," i.e., in a colony of four to six consortia.

The green partner is probably identical to the free-living chlorobacterium *Pelodictyon*. It is able to live independently. However, the consortium is so highly integrated that, for example, the famous expert on lower organisms Lauterborn (1906) described it as a single bacterium—colorless, with green dots.

Only Buder (1914) managed to prove that this mottled bacterium is in fact a system of cytodes, but it is still called by one name, the name of the complex, *Chlorochromatium aggregatum* (or *Chloronium mirabile*).[8]

In another, similar case, the center of a consortium is occupied by a cylindrical, blunt-ended, colorless bacterium without flagella. Inside its mucus are lodged numerous small (approximately 0.33 μm in diameter) globular blue-

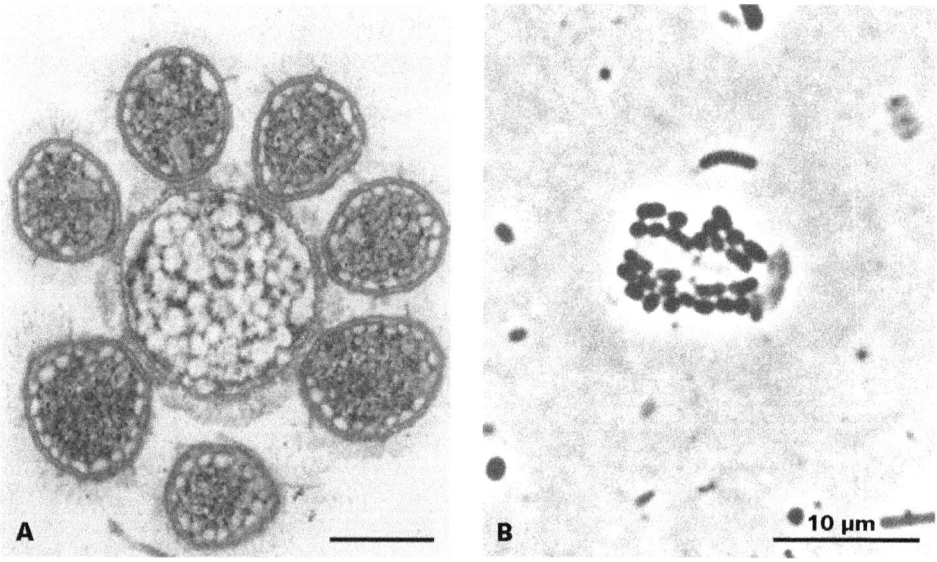

Figure I-1. *"Chlorochromatium aggregatum"*: a phototrophic bacterial consortium. A, Green sulfur bacteria, seven transverse sections, surround the one motile organochemoheterotrophic bacterium. TEM (bar = 1 μm). B, When the live consortium bacterium that appears to be a mulberry-shaped very large bacillus is crushed, the motile heterotroph can be seen as a pale area in the center at the light microscopic level (bar = 10 μm).

greens (or green bacteria). Sometimes their number is so high that the entire consortium appears to be a singe, green cell. The green component is also found as separately living colonies.

When the central bacterium divides, cyanobacteria are inherited with the mucous covering. When spores are formed by the central organism, the green partners attach to the central organism, thus ensuring the inheritance of this symbiotic combination.

The third example (as well as the previous one) was discovered by Pascher [1914]. It differs in having a spirillum, i.e., a corkscrew-shaped rod, 15–18 µm long, in the center of a consortium. A great number of brown-green, ellipsoid cytodes, 1 µm in diameter, are lodged in its mucous covering. Neither partner has been found separately; they probably cannot exist without each other.

A CONSORTIUM BETWEEN A COLORLESS CYTODE AND MULTIPLE PURPLE CYTODES. Its structure is similar to that of *Chlorochromatium,* but instead of green bacteria it has purple sulfur ones. Such a consortium was described by Lauterborn [1906] as a single bacterium named *Pelochromatium roseum.*

This sausage-shaped consortium reached 105 µm in length; its axis is formed by a chain of colorless rod bacteria, i.e., a filament-shaped colony (trichobacteria). This chain is enclosed by a thick mucous covering. A great number of green bacteria, possibly identical to free-living *Pelodictyon,* are located on its periphery. The central filament also has been found without green partners.

This type of consortium was discovered by Perfiliev [1914b], who named the entire system *Cylindrogloea bacterifera.*

Oxygen production during photosynthesis by green or blue-green partners in such consortia may be useful for the central colorless organism (Buder [1914], Pascher [1914]).

Two possibilities exist here. First, the central organism, due to oxygen production by its photosynthetic partners, may be able to thrive in habitats where it is unable to live by itself. Second, green and blue-green consortium partners may facilitate phototaxis of the central bacterium. Green bacteria, and blue-greens themselves, probably consume the mucous covering of their cohabitant (mutualism–commensalism).

Finally, the central flagellum-bearing organism moves its partners, which benefits them in their struggle for existence.

Each of the zoogloea-consortia components are found in fresh water as independent living units. Each consortium unit, in structure and function, represents something incomparably more complex and stable than its partners taken separately. Formation of such zoogloeas can be regarded as an example of evolution through symbiogenesis—origin of more "perfect" organisms from more simple ones through a natural synthesis.

4. Symbiosis of Cytodes with Unicellular Organisms [Protoctists]

The bacterial consortia described above became known very recently. Nobody previously even suspected their existence. There is no doubt that they are found in great numbers everywhere, and so far we have no idea of their diversity.

There are also not very many examples known of symbiosis of cytodes with the simplest cell organisms; in this case the issue of symbiogenesis existence also appeared quite recently.

A consortium composed of colorless flagellates (i.e., unicellular organisms) and blue-greens or green bacteria was described by Pascher [1914]. This consortium is very similar to *Chlorochromatium*, described above. However, the center of the zoogloea is occupied by a flagellate, more or less egg-shaped, with a nucleus and a single flagellum on the pointed end (which signifies a vacuole) but without a chromatophore. Next to its front end, the flagellate has an invagination that marks the ventral side of the body; this invagination might correspond to the mouth opening of other protozoans.

In the mucous mantle of this flagellate lie a high number (up to 300) of elongate-ellipsoid green microbes (2–5 μm long and 0.5 μm wide). Only the pointed end of the flagellate is green due to this ornament.

Cell division of the flagellate, as always in this group, is longitudinal. No animal-type feeding by the flagellate [phagocytosis] has been noticed, although it is colorless. The [storage] product of its metabolism is oil.

This consortium lives in shaded, rotting water, polluted by sludge from sugar factories. The green component is also found separately.

In some of Pascher's [1914] experiments with this consortium, which he called *Oikomonas syncyanotica*,[9] oxygen production by green partners is suggested to play a role in the life of the colorless flagellate. When their environment lacked sufficient oxygen, only those *Oikomonas* specimens with a large number of green partners retained motility. The motility of specimens poor in blue-greens was much weaker. Under normal oxygen conditions, all specimens had similar motility.

The benefits brought to the flagellate by its cohabitants probably are not limited to oxygen delivery. Green partners, most likely, consume the mucus of their colorless partner.

In the flagellate *Cyanomonas,* according to Davis [1894], blue-greens live inside the cell, and therein imitate the chlorophyll organelles [chloroplasts]. The physiology of this symbiosis has not been studied.

In the cytoplasm of an amoeba that belongs to the genus *Pelomyxa,*[10] with an absolute constancy, are found peculiar inclusions, already known to researchers. They were first considered to be crystals. Later, it was proved without doubt (Buchner [1921]) that these inclusions are nothing else but colorless bacteria.

Pelomyxa, which do not have these bacteria, should be excluded from this genus, according to Penard [1902]. The name *Pelomyxa* thus can be considered a name of the consortium consisting of the amoeba with its bacteria.

These bacteria lie mostly in the internal layer of the cytoplasm, and in many species they surround the nucleus, like the "reticulate Golgi apparatus" (about which see below, p. 44). In such cases the nucleus becomes evenly covered, from all sides, with numerous rods. Each of these rods, 10–15 μm long, represents a filamentous colony, a zoogloea of two or three, rarely seven to nine, cylindrical bacteria.

The pattern of distribution inside the cell and the shape and size of colonies and separate bacterial bioblasts vary in different species of *Pelomyxa.*

There is no record that an amoeba digests its bacteria. However, under adverse conditions it expels them along with other excreta.

An exceptional constancy of such a combination of amoebae and bacteria, along with the fact that amoebae do not seem to suffer in any way from the presence of the bacteria, suggests that this is another case of a mutualistic consortium. What kinds of benefits are obtained by its participants from this symbiosis needs to be investigated.

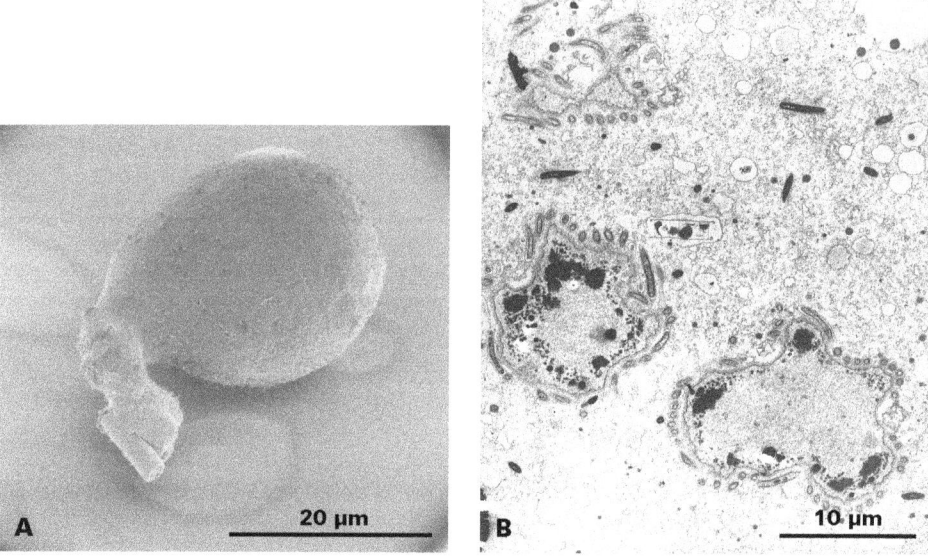

Figure I-2. *Pelomyxa palustris:* a multiple bacterial-protoctist symbiosis. This giant multinucleate amoeba lacks mitochondria, endoplasmic reticulum, and Golgi apparatus but contains three types of symbiotic bacteria. A, SEM micrograph (bar = 20 μm); B, TEM micrograph (bar = 10 μm).

5. Symbiosis of Cytodes with Multicellular Organisms [Animals, Plants, Fungi]

Bacteria and blue-greens have a great propensity and ability to live in intimacy with numerous plants and animals [de Bary, 1879]. They live on cell surfaces, between, or inside cells. As this happens, they form colonies and consortia of great variety.

Blue-greens and green bacteria are consistently found in mucous sheaths of many green and red algae (e.g. *Batrachospermum, Chaetophora,* etc.). Many cytodes, according to Pascher [1914], are not necessarily found in certain species, and therefore are not "obligate." Blue-greens tend to lose color, after which they may switch to saprotrophy. This type of algal-bacterial symbiosis is still not researched.

In mushrooms, according to Baden [1915], spores of the inky cap basidiomycete, *Coprinus sterquilinus,* do not grow without participation of a special bacterium. The mycorrhiza of this mushroom usually is so densely covered by bacterial growth that it cannot be seen. Over four million bacteria may be present on one square millimeter of this mycorrhiza. These bacteria apparently are required for the mushroom's existence, and they do not develop well without it.

Other examples of bacterial symbiosis with plants and animals are listed in chapter III. There we will see how frequently bacteria colonize other organisms, how intimately they interact with them, and what broad formative effect they confer on their associates. We will see below that various cytodes, living inside cells, can imitate plastids, mitochondria, even nuclei themselves and their components, etc.

6. Cytodes [Prokaryotes] as Ancestors

Cytodes (bacteria) can be viewed not only as the simplest but also the earliest of all known organisms. They may be considered as extant representatives of ancestral life-forms on Earth.

Whether we assume that the first organisms were introduced to Earth from other worlds, or that they originated here, we can imagine them only as incredibly small, simply organized, very tolerant to temperature variation, and in general having a great vitality.

This is exactly what the majority of cytodes are.

"Size of many bacteria is close to the size of protein molecules." The dry weight of a very small bacterium, according to Meyer [1920], is 2.7×10^{-13} mg, i.e., 0.00000000000027 mg. The dry weight of a hemoglobin molecule is 1.4×10^{-17} mg.

Meyer even considers it hard to imagine that the smallest bacteria can be made of molecules. In his opinion, they are too small for that. According to this author, their building blocks are "vitules," "something smaller than molecules and consisting not of electrons but of muons, which are 2,000 times smaller than electrons."

"An enormous diversity of bacteria is explained by the fact that protein substances, being the most complex ones, provide an enormous number of variations" (Kolli [1894]). The very appearance of cytodes on Earth is imagined as simultaneous formation of many diverse forms.

Admitting that *noncellular* organisms [cytodes, bacteria] are the original ones, we should pay special attention to details of their biology.

A social way of life, characterized by a preference to form diverse zoogloeas, colonies, and consortia—i.e., systems of variable degrees of complexity—and finally, to invade cells of cellular organisms is highly typical of them.

Zoogloea-consortia, in their architecture and activity, are clearly similar to true cells. In the cases discussed above, the central partner can be compared to a nucleus, and the rest of the covering, including peripheral partners, to the cytoplasm and its inclusions. As we study bacterial consortia, we witness the evolution of the cell.

Another remarkable fact is that bacteria that live as symbionts in true cells of organisms in an astonishing way are similar, topologically as well as often functionally, to the organelles of those cells. Numerous examples are given below.

Study of the symbiotic life of bacteria leads to the recognition that they participated in the origin of cells and to the recognition of their role as cell organelles.

Study of cells as such, as one can see from chapter II (p. 19), leads to the same conclusion.

II

The Cell and Its Organelles

The organization of a true cell, as opposed to that of a bioblast [bacterial cell], can be studied. A cell has constituents compared to organs in a multicellular organism.

Genuine organs are multicellular and consist of tissue combinations. Thus, it is inappropriate to call the cell's parts "organs." A preferred name for cell components is organoids, or organites. [The term "organelles" is used throughout this translation—*Eds.*]

1. Chlorophyll Organelles and Other Plastids

"A living chlorophyll granule represents a marvelous chemical laboratory. With few exceptions, only here does nature make organic materials; from carbon dioxide and water. The help of green pigment [chlorophyll], cytoplasm, and light is required. The first product detected microchemically is starch. It is a point of origin for all other organic substances in plants as well as in animals and humans" (Molisch [1918]).

The exceptions are, e.g., blue-green "algae" and green bacteria, which lack chlorophyll organelles [chloroplasts] but nevertheless perform the chemical work characteristic of them.

(a) Chlorophyll Organelles in Animals [and Protoctists]

A number of green, yellow-green, and green-brown animals that belong to various invertebrate taxa have been known for a long time.[1] This coloration is known to depend on the presence in the animal's cytoplasm of green, yellow-green, etc. organelles or bodies. These bodies are thought to be cell components, and to originate as a product of cytoplasmic differentiation.

Haeckel called these bodies "pigment granules." Others (e.g., Stewart) considered them to be cell nuclei. That these "pigment granules" contained chlorophyll, the substance so characteristic of plants, was discovered by M. Schultze [1851]. Green animals excrete oxygen when exposed to light, and starch accumulates in animal cells that contain "pigment granules."

Most researchers, including some first-rate names, were quite convinced that these pigment organelles originate endogenously. Some tried also to demonstrate that animal "chlorophyll" is not true chlorophyll, but a special, animal "enterochlorophyll."

Cienkowski (1871), a Kharkov University professor, probably was the first to notice that yellow organelles of Radiolaria (unicellular animals [protoctists]) can live also outside of the host cell. They reproduce energetically by division once they are released. They resemble unicellular algae.

The more detailed study led to the conclusion that such "organelles" have cell structure. They are complete cells with cellulose walls, nuclei, and their own chlorophyll organelles (or a single organelle). The "organelles" clearly do not emerge from cytoplasm. They are either inherited or acquired from outside. Several researchers simultaneously concluded that separate, independent plant [protoctist] organisms lead a symbiotic existence inside animal [and protoctist] bodies (G. Entz [1881], Brandt [1881–1883], Geddes [1882], Hamann [1882], etc.). The studies of animals [and protoctists] and their "chlorophyll organelles" are wonderful examples of symbiosis, of consortia between different organisms: animals [and protoctists] and plants [protoctists]. Experimental studies devoted to this issue some time ago showed undoubtedly that the animal "organelles" are in fact unicellular algae.

Yet some outstanding authorities (e.g., Lankester), in spite of the facts, retained the old view and, contrary to evidence, claimed that these algae are cell organelles that originated from cytoplasm. Such is the power of tradition.

Common "chlorophyll organelles" in animals [and protoctists] are small cells (e.g., 1.5 to 10 μm) with a cellulose wall, cytoplasm, a nucleus, and one or a few chromatophores, with or without pyrenoids. Inside these cells are various products of their activity: e.g., oil drops or starch grains.

When an especially intimate cohabitation occurs, such as, for example, in trematode worms, the algal cell wall is very thin and exhibits a chitin rather than cellulose reaction. Or the wall is absent altogether, and the nucleus disappears completely. By fragile appendages, such non-nucleated algae form a connection with neighboring cells of the animal host. Nuclei of the animal cells lie especially close to those algae and appear to substitute for their lost nuclei.

The "chlorophyll organelles" can be subdivided into two groups:[2]

Zoochlorellae: algal cells in the animal that contain one large, cup-shaped, green chromatophore; pyrenoid present.

Zooxanthellae: algal cells contain several parietal, small, yellow chromatophores; pyrenoid absent or present.

Zoochlorellae mainly occur in freshwater animals [and protoctists]; zooxanthellae, in marine animals [and protoctists].

Brandt [1881–1883], who introduced the names "zoochlorella" and "zooxanthella," suggested that "chlorophyll organelles" belong to only two genera of algae. The situation is less simple. Zoochlorellae and zooxanthellae are recruited from various algal genera.

The most common type of zoochlorella, as demonstrated first by Beijerinck (1890), is probably identical to the free-living freshwater and saltwater alga *Chlorella vulgaris* (Protococcales), which has no motile stage.

Zoochlorellae of trematode worms[3] [flatworms *Symsagittifera* that harbor *Platymonas*], when released from the animal's body, develop two pairs of long undulipodia, anterior and lateral. Each has an eyespot and other features. Such zoochlorellae may belong to the chlamydomonads of the *Carteria* type.

Other zoochlorellae have been identified as flagellates: *Sphaerocystis schroeteri* [Chlorophyceae, Tetrasporales]. The systematic position of others is not clear. However, they are not Beijerinck's *Chlorella*.

Common zooxanthellae are flagellates that belong to cryptomonads. Brandt [1881–1883] considered some of them close to the peridinea flagellates [dinomastigotes] with their cellulose armor made of separate plates.

The systematic position of all algae-"chlorophyll organelles" is still not known. However, they are diverse and specific for the animal [or protoctist] group.

These false "chlorophyll organelles" are found in many rhizopods, infusoria [ciliates], and flagellates [mastigotes], and also in sponges, coelenterates, and trematodes [flatworms]. To a much lesser degree the same symbioses are found in annelids, rotifers, bryozoans, echinoderms, and mollusks, where, however, they have just begun to be studied.

From one to several thousand individual zoochlorellae or zooxanthellae can be found in a single unicellular animal [protoctist]. They are usually located in the granular layer of cytoplasm.

If only one "organelle" is present, it divides simultaneously with the animal cell's division. "Organelles" do not hinder the host cell's division. They are inherited in equal quantities by both offspring cells. The alga may outlive its host and return to free living.

Under adverse conditions, the animal cell expels its "organelles" (e.g., when in the dark) or digests them (when starved).

Infection of colorless protozoa by free-living algae, or algae from other individuals, is clearly possible. However, it matters which algae, or "organelles," of which host were selected for experimental infection. The majority of hosts seem to have their own specific tenants.

Animals [and protoctists] become infected through engulfing. If algae are inappropriate, they are simply digested.

Among colorless unicellular animals [protoctists], some are immune to algae, while others harbor algae only facultatively.

The relation between zoochlorellae and free-living algae is indisputable, even though their experimental cultivation after artificial extraction from animals is often unsuccessful. Probably, certain physicochemical conditions are required for the existence of those "organelles." After a number of generations inside the [protoctist or] animal cell, they lose their capability for living freely.

In coelenterates, xanthellae and chlorellae usually populate endodermal cells and are found not only in glands. They are also found, less commonly, in ectoderm. During asexual reproduction, algae are easily inherited. Hydroid polyps inherit algae through eggs, which therefore may be green or have other color. Scyphoid polyps and corals probably acquire larval infection through the oral opening. Digestion of algae has not been confirmed.

How exactly do algae penetrate the eggs of hydra and other classical "organelle"-inheriting animals? This is an important research question. The flow of nutrients directed from endoderm to actively growing and accumulating yolk carries algae as well, suggest Hamann [1882] and others. Algae, stimulated by the nutrients accumulated in the eggs, move on their own, suggest Hadži and others.

"Without a doubt,—writes Buchner [1921]—an assumption of active motility, guided by chemotaxis, under the stimulus produced by the egg, is the most simple explanation for this act, by analogy with the act of fertilization. However, it is not clear how, with their thick cellulose covering, the algae could feel the stimulus and how, being devoid of locomotory organs, they would be able to move on their own" (flagellated cells lack flagella in their intracellular stage).

The assumption that algae are carried passively by the flow of fluid does not explain why transfer occurs in some but not in other cases, as in animals with the same morphology where algae never come into contact with the eggs. The current of fluid penetrates through tissues with no problem, but how chlorellae and xanthellae enter is not known.

The mode of egg infection still must be investigated. However, probably in cases where the symbiosis of algae with eggs in the mother's body is "planned," the tissue is permeable to the symbionts.

In trematodes, zoochlorellae are located mainly in the lacunae of mesenchyme, which fills the space between outer epithelium and digestive tube. Zooxanthellae are found inside the epithelium of the digestive tube.

Contrary to the opinion of Haberlandt [1891] and others who accepted transfer of algae (in colorless mode) via eggs of host worms, others (e.g., Buchner [1921]) maintain that algae infect the cocoon with eggs, or young worms (through oral or genital openings). The cocoon indeed receives a certain number of algae from the mother, but most common infection is likely performed by free-living algae.

We can accept without a doubt that various symbioses of chlorellae and xanthellae with animals are based on mutualism.

Infected protozoan animals [protoctists] develop better under light than those free from green symbionts. They live for a long time without or with less ingested food. The algae develop splendidly during their intracellular lives.

Especially distinct is the mutualistic nature of this relationship and close reciprocal connection of algae and animals in worms, e.g., in the classic dark-green *Convoluta roscoffensis*. Here, algae lose both cell wall and nucleus.

Larvae [of *Convoluta*] that lack algae die. A mature green worm lives for several weeks (*Vortex viridis* [now *Dalyellia viridis*], according to Graff [1891], four to five weeks) without any food. Evidently, it is fed completely by intra-cellular algae. The worm exhibits phototaxis; green tenants are afforded better conditions for photosynthesis.

Here we see a remarkable analogy with the phototaxis of *Chlorochromatium consortium* (see p. 12), phototropism of leaves of green plants, and other phototropisms.

When a worm is killed or dies on its own, its algae die with it. The worms kept in the dark exhibit symptoms of starvation, and may die.

What kinds of advantages do partners in such consortia gain?

It has been experimentally confirmed that:

1. The alga incorporates carbon dioxide and ammonia excreted by the ani-mal. The ammonia is used as a source of nitrogen.

2. (a) The animal feeds on extra starch of the alga. In dissolved form, starch diffuses through the algal cell wall into animal cytoplasm; (b) oxygen, excreted by the alga during photosynthesis, is respired by the animal; (c) through its consumption of ammonia, the alga relieves the animal from its excreta. Young worms, prior to acquisition of algae, die due to poisoning by metabolic products. In their cells vacuoles con-tain long, sharp crystals, absent in green individuals, which probably contain urea; (d) the worm digests part of the reproducing algal popu-lation in some cases.

"Chlorophyll organelles" in animals show us that extraneous, autonomous, and highly organized organisms can inhabit cells just as do organelles. These alien organisms, considered by scientists to be products of cytoplasmic dif-ferentiation, were organisms, now organelles, which invaded cells from out-side. This evidently true understanding is now so common that it is found even in elementary textbooks. However, it came to science late and won ac-ceptance only with great effort.

Physiologically, such acquired organisms play the same role in colorless animal cells as chlorophyll organelles [chloroplasts] do in plants. They were misidentified by previous researchers.

(b) Chlorophyll Organelles in Plants [and Protoctists]

As opposed to green or blue-green cytodes [bacteria], where chlorophyll is distributed throughout the cytoplasm, and to green animals, where the pigment is concentrated in cells of their chlorellae or xanthellae—in all true plants (excluding fungi and a few colorless, parasitic flowering plants)—chlorophyll is localized exclusively in the specialized bodies or organelles of the cytoplasm called chloroplasts or green chromatophores.[4]

Chloroplasts, located in the cytoplasm, are denser than the surrounding cytoplasm and are therefore distinguishable even when they lack color.

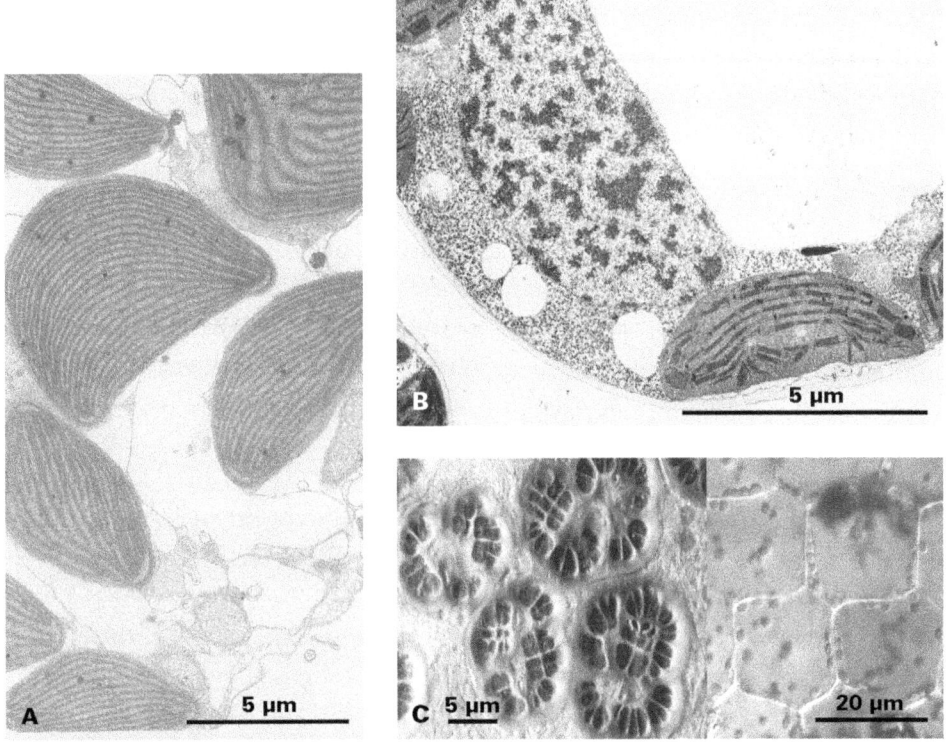

Figure II-1. Chloroplasts: photosynthetic endosymbiotic organelles of cyanobacterial origin in protoctists and plants. A, chloroplasts in algae *(Vacuolaria)*, TEM micrograph (bar = 5 μm). B, chloroplasts in barley *(Hordeum)*, TEM micrograph (bar = 5 μm). C, side-by-side images of cyanobacteria *(Gomphosphaeria)* (left) and chloroplasts of hornwort *(Ceratophyllum)* (right), light microscopy (bars = left, 5 μm; right, 20 μm).

The most common type of chloroplast, found in all plants, is the so-called chlorophyll organelle. Spherical, lens-shaped, ellipsoid, or, when densely packed, multifaceted, its number in a single cell varies from one to many. Its average size is 5 μm.

The inner structure of an organelles's body (stroma) appears "spongy," reticulate, or granular. It is comparable to the structure of cytoplasm in general, and to bacterial bioblasts [cells] in particular. Sometimes chlorophyll is distributed in spherical granules dispersed throughout the entire stroma. These are oil drops in which chlorophyll is dissolved (Pringsheim [1915]). Such granules resemble those of chromatoplasm in blue-green "algae." Other, mainly enigmatic "grana" are also encountered, which are comparable to grana found in cytode bioblasts [bacterial cells].

A special covering of stroma, a peristromium, was noted by Senn [1908] and others; other authors disagree.

Chlorophyll organelles [chloroplasts] are very important parts of plant cells. No gradual transition between diffuse, dispersed distribution of chlorophyll characteristic of blue-green cytodes, and chlorophyll organelles, sharply distinguished from the rest of the cytoplasm in lower plants [algae], is known. On the contrary, lower plants [algae], including unicells, not only possess sharply distinguished chloroplasts but are usually even more complex than chlorophyll organelles in higher [flowering] plants. They often have pyrenoids—colorless proteinous bodies of an enigmatic function. We have seen already that, in chlorellae, the chloroplast is positioned, like a cup or a hat, on top of such a pyrenoid.

Morphologically, chlorophyll-bearing bioblasts of the cytodes [bacterial cells] are comparable, in their entirety, to a single chlorophyll organelle. But in no way do they resemble cells that contain organelles such as nuclei and other structures.

Therefore, a very large gap exists between the type of chlorophyll distribution in cytodes, on the one hand, and in plants composed of cells, on the other. We see no hint of gradual transition. It is very hard to imagine a morphological path of differentiation in which such transition [from cytode to plant cell] could happen.

If one assumes that chloroplasts were a product of differentiation of the cytoplasm that originally lacked chlorophyll organelles, then the physiological transition is obscure. From the physiological viewpoint, what caused such differentiation? There seems to be no benefit in the existence of chloroplasts relative to diffuse chlorophyll distribution.

Scientists were convinced for years—just as in the case of chlorellae and xanthellae—that chloroplasts were granules of dye produced by the cytoplasm (A. de Candolle). They were products that originated *de novo* from cytoplasm every time (Unger 1846, Mohl 1851, etc.). But in the early 1880s Schimper, Schmitz, and Meyer proved that this is not true.

Chloroplasts are never produced by the cytoplasm. They always originate through division, one from another, and are inherited like hydra's chlorellae. They are present in eggs and seeds, and inherited only from the female side, just like, in many cases, chlorellae and xanthellae.

The inheritance through spermatozoa is also very peculiar. In *Funaria* moss, a mature spermatozoon that bears two flagella [undulipodia] consists of a more or less filamentous nucleus on its pointed end. A rounded chloroplast attaches to the opposite end and is "delivered" by the motile spermatozoon to its destination. This process resembles the relationship of two partners in *Chlorochromatium* and other consortia.

Chloroplasts were claimed to be absent from the reproductive organs, in cases cited as evidence against the theory of Schimper [1885] and Schmitz [1883]. Then the existence of chloroplasts as smaller, colorless units was confirmed. Such units were often misidentified as mitochondria. However, some mitochondria may possibly give rise to chloroplasts.

Chloroplasts thus may originate from colorless plastids. They also can give rise to non-green colored plastids (chromoplasts) or colorless plastids (leucoplasts). Chromoplasts differ from chloroplasts not only in their pigment but also in morphology. The nature of chloroplast derivatives will be solved together with the nature of chloroplasts.

Lyubimenko [1916] summarized our modern views: "The genetic nature of chloroplasts at all stages of plant development is clear. If they by some reason disappear from a cell, the cell cannot form them again."

The physiological independence of plastids is also very high. In this sense they strikingly resemble chlorellae and xanthellae.

They reproduce by division, like cytodes and lower animals [protoctists] and plants [algae]. They exhibit pairwise fusion, which resembles a sexual act of protozoa. They exhibit reactions similar to those of unicellular animals [protoctists] such as rhizopods and radiolarians. Their response to light is the same as in green cytodes. They move on their own, sometimes by pseudopodia—the cytoplasmic outgrowths and lobes—choosing their path independently, according to Weiss [1864–1866], Senn (1908), Küster (1911) and others.

Their independence from the rest of the cell is seen in their relationships with nutrients and oxygen.

The light sensitivity of plastids, according to Lyubimenko [1916], is based not on the formation of carbohydrates, utilized by the rest of the cell, but on the formation of nitrogenous compounds necessary for plastid nutrition. Accumulation of the chlorophyll, previously considered the main function of chloroplasts, is the result of an organic nutritional process of the organism, and performed independently of photosynthesis, even in the absence of light. Chlorophyll is, possibly, not more than an excretion of excessive nitrogenous nutrition.

The plastids produce a special antienzyme, a ferment that blocks chlorophyll oxidation and limits its role as a "universal sensibilizer."

Formation of starch is a physiological reaction of plastids to certain nonnitrogenous compounds when the concentration of those compounds attains a certain high level. We have no data that plants need carbohydrate accumulation in the form of starch. For plastids, starch formation is protection from excess carbohydrate.

The plastids clearly prefer organic nitrogen, while colorless parts of the cell use nitrogen from nitrates. Glucose, fructose, and maltose are utilized only by colorless cytoplasm, while xylose, dulcite [isomer of mannite], etc., and also glycerin and lactose are utilized by plastids.

The cytoplasm requires oxygen. However, the absence of fast oxidation of chlorophyll during photosynthesis indicates that "the atmosphere for plastids inside the cell appears to lack oxygen."

Under the action of hydrofluoric acid, the colorless part of the cell was destroyed, while chloroplasts stayed intact, in the famous experiment by Fischer.

Reinke observed the following in 1870s. In a rotten pumpkin, after the colorless parts of cells were completely destroyed, the chlorophyll organelles continued to live and reproduce, when surrounded by the mycorrhiza of *Pleospora,* an ascomycete. With this fungus they formed a consortium like that of lichens (see below).

Convinced that the chlorophyll organelles are equivalent to chlorellae and xanthellae, A. S. Famintsyn already in 1868 attempted to obtain a pure culture of chloroplasts.

To extract chloroplasts from cytoplasm, he suggested two methods: (1) paralysis of one cell part to maintain development of another; and (2) lysis of the cell to find the conditions for culturing extracted chloroplasts. Famintsyn himself worked on the second experiment but did not obtain the desired results.

Much more successful were other experiments by Lyubimenko (1918). Based on the different relationship of plastids from the rest of the cell to oxygen, he suggested a tissue be placed in an almost oxygen-free atmosphere where the plastids would outlive both cytoplasm and nucleus. He thought it would be possible to isolate plastids from other living components without killing them.

Fragments of leaves from various plants were placed in vials with water in a layer 1 cm thick at room temperature. In one to two days the decomposing bacteria developed in the leaf mass; in three to four days after that the leaf fragments sank to the bottom. In the dead fragments, cytoplasm and nuclei were dissolved, while plastids remained undamaged and mainly retained their original color. After that cell walls dissolved. Approximately two weeks later, only the most stable parts from the tissue remained: thickened parts of vessels, mechanical fibers, etc.

But plastids still retained their appearance and color, as they were submerged into a mucous mass formed from dissolved cell walls and slime secreted by the bacteria.

Later, the isolated chlorophyll organelles were cultured in various media. Dividing and accumulating plastids were observed, apparently indicating that reproduction had taken place. However, this was not possible to prove. Division of plastids ceases when the cell growth stops. No starch formation, in glucose-based cultures, was detected. However, plastids in living tissue under the action of acids also lose their ability to deposit starch. We also note that luminescent bacteria in pure culture often lose their ability to luminesce.

Plastids demonstrated very different reactions to different nutritive media; in some they exhibited symptoms of starvation; in others they died, but removal of atmospheric oxygen did help to preserve plastids.

The flora [microbiota] that accompanied plastids in these experiments was exclusively bacterial. Plastids resisted the action of decomposing bacteria.

The stroma of isolated plastids was stained by certain dyes absolutely in the same way as the plastid stroma in the living plant.

Therefore, for an impartial observer, the isolated plastids appear as autonomous organisms that live in cells much like chlorellae and xanthellae.

Observing the mode of plastid inheritance, Schimper (1885) asked whether they are not algae that live symbiotically in naturally colorless cells. However, science knows no such algae.

A possibility that chloroplasts originated from unicellular algae such as chlorellae by simplification, or degeneration, was discussed by Haberlandt (1891). Such simplification seems to be indicated by the above-mentioned loss of nucleus and cell wall by chlorellae in green worms.

No traces of such origin are seen in true chloroplasts and in plastids in general. If one assumes that this was the case in most multicellular plants, the question remains: what are the chloroplasts of those unicellular algae (chlorellae, etc.) from which the true plastids originated, and how were they formed? That is, under Haberlandt's hypothesis we are still left with the open issue of plastid origin.

That plastids are blue-green cytodes, Cyanophyceae, which have adapted to intracellular existence, Merezhkovsky (1905) tried to prove. In his opinion, pyrenoid of chloroplasts corresponds to the "central body" of blue-greens, while hat-shaped (as in many algae) chloroplast corresponds to their chromatoplasm. It is important to remember that Merezhkovsky's conclusions were based on Fischer's [1897] description of blue-greens' structure, which, as far as we can judge today, is incorrect.

However, there are no known green plastids that contain phycocyanin, and equivalency of the pyrenoid and the "central body" is highly problematic.

The reasons to consider chloroplasts as a type of green cytodes from the same group that is gradually becoming known as "green bacteria" are several.

Whether similar cytodes may be discovered in a free-living condition, we can only answer by noting that until very recently nothing was known about green bacteria and remarkable bacterial consortia. Nobody suspected that chlorellae and xanthellae could exist as free-living organisms.

A group related to green plastids, with their modifications such as leucoplasts, chromoplasts, and false mitochondria, are red, brown, and yellow plastids found in various groups of algae. Chlorophyll in these species is masked by other pigments. These organelles in principle should not differ from chloroplasts. One can apply to them the same conclusion that was reached in chloroplast studies, that they cannot originate by cytoplasmic differentiation and are inherited during reproduction through division.

If one takes only the color of plastids as the basis for systematic division of algae, the resulting groups will be in many cases so diverse as if somebody started to group lower animals based on the color of their symbiotic chlorellae and xanthellae (Merezhkovsky [1905]).

The red "eyespot" of flagellates [mastigotes] and algal spermatozoa was considered a modified chloroplast by some researchers (e.g., Rotert [1891]). However, the continuity of its hereditary transmission, the sharp difference from chloroplasts, etc., as well as the absence of gradual transitions between the two, most likely indicate that this is a special type of symbiont, "organelle," that originated also from special mitochondria, although related to chloroplasts.

2. Centrosomes

The centrosome, a characteristic organelle of animal cells,[5] was considered, e.g., by N. Bernard [1916] as the first difference between animals and plants. Earlier it was suggested that it is also present in plant cells (e.g., Guinniard, Strasburger). Today, after Tischler [1922], it is accepted that something similar is found only in algae. The centrosome in plants may be closely related to the so-called blepharoplast (Sharp [1914]), discussed further below.

A typical centrosome usually consists of two or more centrioles—specific small bodies—and a sphere surrounding them. The inner layer of the sphere differs distinctly from the rest of it and forms a spherelike dense body (Boveri's centrosome), which contains centrioles.

The peripheral portion of the sphere has a structure that changes with age. Most likely it is just a result of specific distribution of the parts of surrounding cytoplasm rather than a special organelle.

The nature of centrosomes and centrioles is enigmatic. Centrosomes, as a rule, play a role in cell division. Dividing in two, a centrosome forms two "poles" of the karyokinetic [mitotic] spindle; it apparently triggers and guides nuclear division.

To understand the relationship between centrosomes and other, somewhat better-studied cell organelles, the following fact reported by Němec [1910] is important.

The cells of a hornwort *Anthoceros* have a spherical nucleus and one ellipsoid or spherical chloroplast with a pyrenoid. The chloroplast lies close to the nucleus. Prior to nuclear division, the chloroplast divides by fission. "After this division in two, its halves move apart and are located at the poles of the future formation of division." Later, the spindle moves right between the halves of the chloroplast.

"This process resembles so much the behavior of animal cell centrosomes that the chloroplasts could be recognized as centrosomes if they did not contain chlorophyll and starch" (Němec [1910]).[6]

Regular chloroplasts can become colorless and sometimes lack starch. One has to accept that the chloroplast of *Anthoceros* represents both a chloroplast and, in fact, a centrosome, alternating those two functions.

Animal centrosomes may also be a special kind of plastid, i.e., bodies of the same nature as chloroplasts, i.e., cytodes.

3. Nuclei

In some species of the filamentous green alga *Spirogyra,* the organelle that was considered its regular nucleus is in fact an entire cell itself, with its own membrane, cytoplasm, etc., and with a real nucleus inside (Famintsyn [1907, 1912]). This cell is an autonomous unicellular organism, which lives symbiotically in *Spirogyra*'s cell (Famintsyn [1907, 1912]). Famintsyn isolated it from the cell and tried to cultivate it separately. These types of nuclei are also present in some gregarines, radiolarians, and rhizopods, where, as pointed out already by Carnoy (1884), nuclear karyokinesis takes place not in the "nucleus" but in the "nucleolus." Nuclear staining highlights the "nucleolus" rather than the "nucleus."

So-called yolk nuclei [vitelline bodies] are characteristic for the eggs of spiders and some other animals. Many authors considered them real nuclei; e.g., Balbiani [1864] thought that they originate through budding from regular cell nuclei. Under pressure, the content of the yolk nucleus comes out from its membrane as numerous rodlike bodies.

That this type of nucleus is, without a doubt, a colony (or consortium) of bacteria was confirmed by Portier (1918), who succeeded in culturing them in situ. When the membrane disappeared, these bodies came out into the surrounding cytoplasm and filled the entire egg capsule.

However, the yolk nucleus is similar, for example, to the macronucleus of infusoria [ciliates], which appears to be a "doubtless" true nucleus. One cannot exclude that the bacterial nature could be proved in this type of nucleus as well (Portier 1918).

The cell "organelles" in luminescent tunicates, claimed to be true nuclei by A. Kowalewsky, proved to be accumulations of bacteria according to the most recent studies of Pierantoni [1913] and Buchner [1921] (see below).

A number of such examples show that an extraneous organism, or an accumulation of such organisms, may be mistaken for cell nuclei. The best authors did not suspect for a long time that they mistook bacteria for nuclei.

Those "nuclei," which until now are considered "doubtless" cellular organelles, in various organisms and in various tissues "differ so much in their structure and processes going on in them, that one can indicate only a single feature common to all of them. This single character changes very little, and it can be formulated as follows: a nucleus is a certain structure that lies inside a cell. It usually is sharply distinct from the rest of the cell's contents. Thus it appears unreasonable to group all such structures in one category and consider all of them equivalent" (Famintsyn [1907]).

We discuss here only typical, "normal" nuclei, the kind described in all biological textbooks.

The nucleus was long touted to be a result of differentiation of cytoplasm. This opinion is proved wrong: a nucleus always originates from another nucleus. Nuclei form only by division of parent nuclei.

A great number of facts confirm an "uninterrupted individuality" of chromosomes, those most important nucleus-forming structures. Chromosomes originate from other chromosomes through reproduction by division. An entire nucleus can be formed by reproduction of a single chromosome, according to Boveri [1887].

An opinion exists that chromioles—granules of chromatin carried by a colorless chromosome body—also reproduce, each separately, by division, and also represent inherited cell structures.

With the absence of evidence that the nucleus originates from surrounding cytoplasm, one must note from the following facts a great independence of the nucleus from its cell, its unexpected autonomy for an organelle.[7]

We describe a case of symbiosis between an orchid and a fungus. The fungal mycorrhiza occupies a number of cells of the orchid. As it grows, it begins to fill, in a globular shape, almost the entire cell; meanwhile, the nucleus becomes compressed in the center. Then the nucleus suddenly makes its way, wiggling, to the closest cell wall. It moves actively, resembling an amoeba, and on its way falls into pieces, which independently make their way to the wall. There, next to the wall, the nucleus restores its integrity and "obtains a certain, almost mysterious acting force" (Burgeff [1909]). The results are soon seen. The outlines of mycorrhiza become less definite; its hyphae fuse and turn into shapeless masses and are slowly digested.

A significant number of examples of "migration" of nuclei from their cells are known, when they crawl from cell to cell. When a binucleate cell is formed in rust fungi this is not exceptional or pathological. A nucleus crawls into the neighboring cell of the same hypha, and also into a cell of another hypha if the cell lies close enough. The nucleus assumes a filamentous shape and squeezes itself into a very narrow cell wall pore. A nucleus that leaves its cell does not return; rather, it lives from then on in an alien cell. The abandoned cell slowly loses its cytoplasm and empties [Welsford 1915]. Another such example is known during apogamy in ferns [Bower 1923].

In green worms, chlorellae eventually lose their nuclei (see p. 21). However, the nuclei of the neighboring cells of the worm approach the algae and functionally substitute for their lost nuclei (Buchner [1921]). A nucleus of an animal cell may take the place of a plant cell's nucleus.

The eggs of echinoderm animals, sea urchins and sea stars, can be forced to fall into fragments, some of which lack nuclei. When a spermatozoon—a cell that is considered to be an almost naked, or completely naked, nucleus—is introduced to such a fragment of cytoplasm, a viable cell results. In the experiments of A. Kowalewsky and Boveri, the nuclei of tunicate spermatozoa were introduced into the bodies of mollusks, where they entered the cells of their spleen. They retained viability, reproduced in the alien cytoplasm, and existed in it in the same fashion as its own nuclei.

These observations do not correspond to the idea of a nucleus as an unalienable part of the cell, an "organ" that exists in its interests. Once again they bring to mind zoochlorellae and zooxanthellae, *Spirogyra*'s nucleus, "yolk nuclei," and so on.

At the same time, the role of the cell nucleus is still completely unknown. Functions of this organelle are still extremely problematic, and physiologically the nucleus is a structure "enigmatic to a high degree" (Borodin [1910]).

It is thought that the "nucleus should play the most important role in metabolism and this role lies in its structural, assimilatory portion, in its chemical and morphological synthetic processes and secretion of various complex products, e.g., cell wall" (Maksimov [1914]).

All this is, to a certain degree, a conjecture. There is very little data to confirm such a role. All cells have nuclei, and cells lacking nuclei are not viable. The nucleus can approach the site of nutrient intake, form special processes and lobes toward it, etc. The nucleus is located in the area of a cell's maximal growth; cell wall of a plant cell grows especially well at the side where the nucleus is located. By its division, the nucleus triggers the division of the entire cell.

However, as we address these issues it is necessary to note the following:

The nucleus cannot be considered absolutely necessary for life already, because numerous non-nucleated organisms have exceptional vitality. Areas of cytoplasm from which nuclei are artificially removed remain alive for a shorter time, but they also react to stimuli, perform photosynthesis, etc. Also, the process of cell respiration likely proceeds in a regular way: at least, "when air is substituted by an indifferent gas such as hydrogen or nitrogen, the fragments containing a nucleus die as fast as those without it" (Maksimov [1914]).

Movement of the nucleus toward nutrients shows that the nucleus itself is interested in them, but this hardly can be a confirmation of its special role for the benefit of the entire cell.

The stimulating effect of the nucleus on cell wall growth cannot be considered to have been confirmed even in plant cells, while in a great number of cases cells remain naked, the presence of a nucleus notwithstanding.

In many cases nuclear division does not trigger cell division [cytokinesis]; bi- or multinucleated cells are produced, or division begins not with the nucleus but with other cell organelles (e.g., in some hornworts, with chloroplasts, see p. 32).

Algae, as discussed above, or fungi and bacteria, as will be discussed below, when they settle in alien cells, definitely assume "a significant role in the metabolism." Such symbiotic organisms usually exhibit chemotaxis, stimulate growth of the cell, in particular of its wall. As we have seen, worm cells that lack their symbiotic algae cannot lead a normal life and are destined to die.

In bacterial consortia such as *Chlorochromatium,* division of a central symbiont, analogous to a nucleus, leads to the division of the entire zoogloea.

The features of the nuclei, therefore, are also exhibited by autonomous organisms, in particular by cell-inhabiting symbionts.

The nucleus has been touted to be a carrier of "hereditary substances," although this belief was weakly supported. Heredity is present in non-nucleated organisms as well. However, some features are clearly inherited without participation of the nucleus. Chloroplasts, such important parts of a plant, are inherited independent of a nucleus. In a cross of white-leaved and green-leaved forms of *Pelargonium,* one can trace transmission of white plastids to the offspring from the female plant, and green from the male plant (Baur [1909]). Their combination determines the diagnostic mixed color of the hybrid's leaves.

We should allow that nuclei "possibly do carry a hereditary substance from parents to their offspring, but that cytoplasm (with its inclusions) has the same ability, and currently it cannot yet be decided which task in hereditary phenomena belongs to either part of the cell" (Stieve 1923).

But the nucleus has no monopoly as a carrier of hereditary units.

Therefore, the role of the nucleus as a cell organelle is highly enigmatic, and its suggested physiological features are clearly not exclusive. From the roles ascribed to the nucleus, we still do not understand characteristic features of its structure.

The best analogy to nuclear structure in organic nature in general is the bacterial zoogloea. The cell nucleus is nothing else but a collection of bacteria in close proximity—symbionts of a cell that is non-nucleated, suggested Boveri [1887].

Such bacterial colonies and consortia, which lead a mutualistic, intracellular way of life, are known in insects. The symbionts of a leaf louse [a coccid], *Pseudococcus,* were studied by Pierantoni [1913] and Buchner [1912, 1921]. Bacterial colonies are represented by spherical bodies, several of which are located in each cell of a special organ (see p. 82). These spheres are about the same size as the nucleus of the same cell. Just as the existence of nuclear envelope is mysterious, no final opinion exists on the presence of such an envelope in bacteria. Pierantoni [1913] claimed that it exists, and Buchner [1912, 1921] that it does not exist.[8]

These spheres contain "mucus" comparable to nuclear fluid, a variable number of cytodes comparable to chromosomes, and enigmatic, structureless granules comparable to nucleoli. At the first stages of colony development, bacteria are diverse in their shapes and sizes. They either are collected into a compact globule, where details are hard to distinguish, or fall into individuals. These individuals may be nearly spherical, rod-shaped, horseshoe-shaped, etc. Sometimes they are supplied with extensions like the chromosomes in a nucleus at its resting stage [interphase].

These colorless rod-shaped and horseshoe-shaped structures are peppered with extremely tiny spots that sometimes form groups. They are differently stained and may be compared to chromioles. Some rods are also known that exhibit alternating, beadlike areas located one above another; they resemble the structure of chromosomes made of so-called chromomeres.

Inclusions, possibly like the nucleoli of true nuclei, are collections of excreta or stored substances.

Similar inclusions to these resemble nuclei in their response to specific stains. Staining patterns of the substance of bacteria that resemble "nuclear substance" also deserve attention.

Such bacterial forms like twins of nuclei exist [and probably] play an important, if enigmatic, role in cell metabolism. They are inherited through the egg.

Chromosomes are "elements or individuals, so to say, most elementary organisms, which lead their independent existence in the cell," wrote Boveri [1887].

Objections have been stated against his "bacterial hypothesis."

(1) The number of chromosomes in a nucleus is constant for each species (only germ cells have a halved number compared to somatic ones); (2) chromosomes can be unequal (heterochromosomes), and their given set is inherited; at the same time, bacterial zoogloeas consist of same elements; (3) chromosomes are able to fall into smaller viable elements, maybe into chromomeres or chromioles; and, when chromosomes are restored by the time of nuclear division, those units could be to some extent redistributed; (4) chromosomes divide lengthwise, while bacteria divide by transverse fission.

These objections are not, however, be taken too seriously. The number of chromosomes is not as constant as it is thought (Strasburger, Della Valle, and others). Not only can it change within a species (*Salvinia natans,* a water fern, has from eight to forty-eight chromosomes), but it is not absolutely fixed even within a single organism [Litardière 1921]

The constant number of colony members already has been noted in colonies that include bacterial consortia. The number of chlorellae and xanthellae per cell, at least in unicellular organisms, is likely also constant. The number of chloroplasts often (if not always) is constant, and in reproductive cells it is twice reduced compared to somatic ones. Species can even be diagnosed by the number of chloroplasts per cell in the same way as by chromosomal number. The plant *Peperomia metallica* has only four chloroplasts in somatic cells, *P. saundersii* has eight, and *P. arifolia* has twelve (Schürhoff 1908).

A regular distribution of chromosomes during the indirect nuclear division allows many "discrepancies," while chloroplasts also take certain positions during division, as if they were mimicking the chromosomes.

The zoogloea of insect cells, as described (p. 36), has homogeneous components. The widespread phenomenon of regular and inherited association of several different cytodes in one cell must be taken into account. The same zoogloea, or a single host cell, can contain two to five (and possibly even more) kinds of symbiotic bacteria. Cytodes, which reproduce by lengthwise division, have been recently discovered.

The ability of chromosomes to fall apart, "ungroup," etc. indeed allows thinking that an individual chromosome is not an individual unit in the full sense of the word. Such a fact, however, is easily reconciled with the bacterial hypothesis of nucleus origin if we accept that each chromosome is a colony or a consortium, i.e., an individual of a higher order, physiologically indivisible, and from the genetic viewpoint representing a system of individuals of lower order. Such a hypothetical assumption makes it easier to understand various perturbances that often happen in the nuclei (which we cannot discuss here) in the spirit of Boveri's hypothesis. A model for a chromosome can be seen in the "individuals" of *Chlorochromatium* or *Pelochromatium*, which were described above (pp. 12–13).

We mention also facts that could serve to confirm the colonial or consortial characters as well as the bacterial nature of the nucleus.

"Chromidia" was the name first given by R. Hertwig (1902) to small bodies that stained like chromatin, and were first found in the cytoplasm of some radiolaria.[9] Often located in "filaments," they originate from the nucleus, consist of nuclear substance, and may play an important role in a cell's life. The nature of chromidia, which may be regarded as living partners of the nucleus, is exemplified:

In the histology of annelid worms (the entire group) are known special granular and rod-shaped structures, which are found in connective tissue and especially accumulate in places where it forms.

They were recognized as symbiotic bacteria by Cerfontaine (1890), who discovered them. Cuenot (1898) also confirmed their bacterial nature. He found that these "bodies" leave cells of the connective tissue. They are engulfed and digested by other cells, which play the role of amoebocytes.

The bacterial nature of those "bodies" is also accepted by G. Schneider, and probably Buchner.

The most recent researcher of those "bodies" (Trojan 1919) even more interestingly considers it proven fact that they originate from nuclei and are equivalent to chromidia. This was not denied by Buchner (1921).

One must agree that to explain the remarkable similarity of structure and activity of a nucleus to a bacterial consortium, one must allow that the nucleus is, in fact, such a "bacterial globule," which leads its life inside the cytoplasm just like chlorellae, xanthellae, and the bacterial "globes" of *Pseudococcus*.

4. Mitochondria

Mitochondria[10] (chondriosomes, plastosomes), are grainlike or filamentous small bodies, with indiscernible structure, located in the cytoplasm of both plant and animal cells. They are observed in the living cell and after special treatment and staining.

As in bacterial morphology, various forms are distinguished: granule-like, or mitochondria; rodlike, or chondrioconts; filamentous, or chondriomites; and comma-shaped, or vibrioid.

In addition to various mitochondria that are colorless, shiny, and cartilage-like in appearance, can be distinguished chromochondria, which "form pigment inside that is dispersed all over the body without changing its shape or chemical nature" (Prenant 1913).

"Grainlike forms,"—writes Guilliermond (1914)—"may lead to confusion, since there are many grainlike cellular structures that are not mitochondria; but a rodlike form is highly characteristic."

Mitochondria are distributed separately as well as in mainly filamentous colonies and tend especially to concentrate around the nucleus.

The majority of researchers agree that chemically mitochondria are of albuminoid [proteinaceous] nature, with impregnation by lipoids.

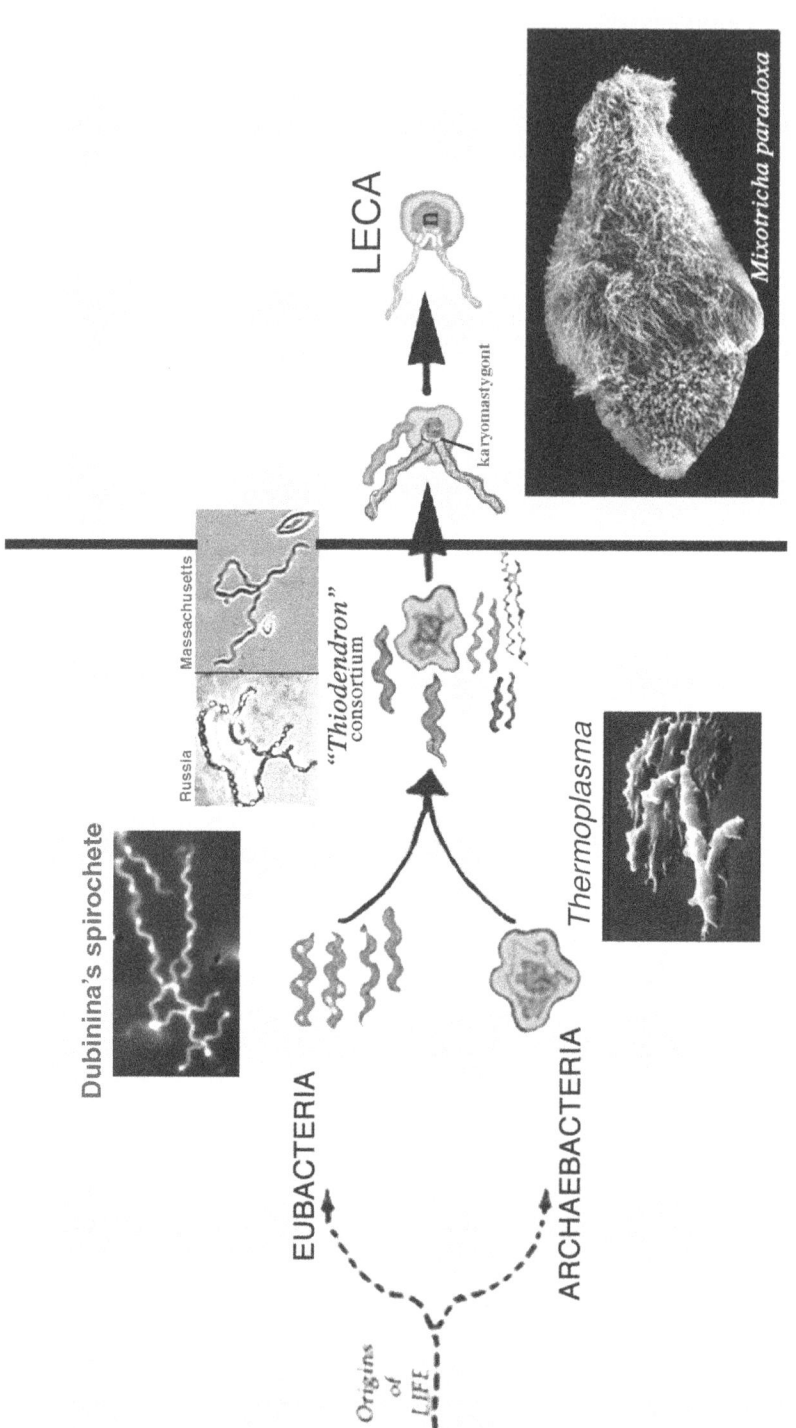

Figure II-2. Eukaryosis: hypothetical eukaryotic cell and tethered nucleus origin (karyomastigont model, eubacterial [spirochete]–archaebacterial [e.g., *Thermoplasma*] symbiogenesis). LECA, Last Eukaryotic Common Ancestor (as in Margulis et al. 2006).

Mitochondria, as "cell organelles," are assumed, of course, to have an "important" but "enigmatic" role in a cell's life. The "chondriome," i.e., the set of mitochondria in a cell, along with chromosomes, according to Meves [1918], represents the apparatus of heredity. This opinion is in agreement with those who specialize in the issues of heredity (e.g., Stieve [1923]). Mitochondria allegedly form tannin and serve for the defense of cells from infection by parasitic fungi, according to Politis [1921].

Some corpuscles or organelles classified as mitochondria are embryonic plastids, according to Guilliermond [1914] and others. They are "proplastids," i.e., precursors of plastids (Randolph [1922]). Others are convinced that proplastids always can be distinguished from "true" mitochondria (e.g., Noack [1921]).

"True" or "permanent" mitochondria are "morphologically and physiologically identical" in plant and animal cells, according to Guilliermond [1914] and others. This statement obviously should be taken *cum grano salis* since, in fact, the structure of mitochondria is not known, and their physiology is quite enigmatic.

Mitochondria reproduce by transverse division. They are inherited, most authoritative experts agree, just like plastids, nuclei, chromosomes, zoochlorellae, zooxanthellae, and other cell organelles. They are able to move on their own (Romieu 1911).

The unusual resemblance between corpuscles, later called mitochondria, and bacteria, was already noted by early authors, e.g., Wigand [1887, 1888]. An opinion was voiced that they are true symbiotic bacteria.

The best modern cytologists, armed with modern microscopic technology and methods of magnification and staining, agree that mitochondria highly resemble simple bacteria. "At some stages, mitochondria have a remarkable resemblance to bacteria" (Fauré-Fremiet [1910]). "If we did not know that these are parts of a cell, we would consider them bacteria" (Mercier [1906, 1911]).

Mitochondria and bacteria resemble each other especially in cases when the latter lead an intracellular life, as in the legume nodules [of *Rhizobium*] or "proteinous" leaf glands of Myrsinaceae and Rubiaceae. Nodular bacteria stain with typical mitochondrial dyes (e.g., by the method of Regaud [1919]) and can be mistaken for mitochondria even by expert cytologists (Portier 1918).

Structures mistaken for mitochondria were found really to be symbiotic bacteria through successful experiments with isolation and growth in pure culture.

The autonomous, true cytode nature was proved for the bodies that were considered mitochondria in cockroaches by Schneider, in some beetles and ants by Strindberg [1913], and in cephalopods by Dubois [1914], for example.

Such a serious cytologist as Meves [1918] stated that "currently, it appears highly possible that mitochondria are nothing else but highly adapted symbiotic bacteria."

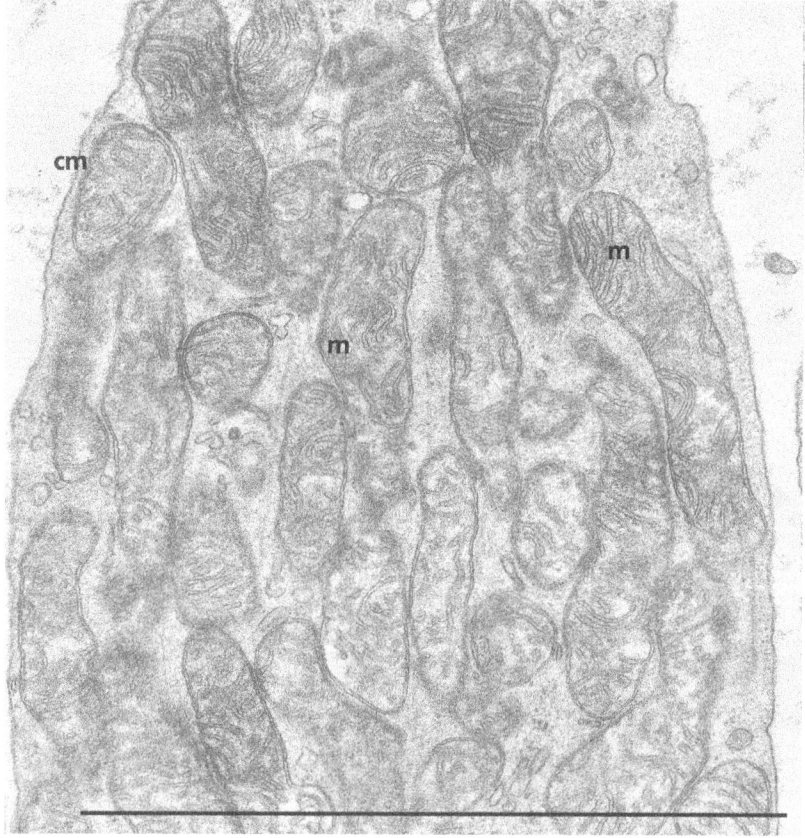

Figure II-3. Mitochondria: oxygen-respiratory organelles of proteobacterial origin, from a mammalian tissue that rapidly consumes oxygen. TEM micrograph, (bar = 10 μm).

This same conclusion was independently reached, through experiments, by Portier (1918). He proved the bacterial nature of mitochondria in his experiments, which included their extraction and placement in culture, just as it is done to simple bacteria. According to Portier, he succeeded in isolating mitochondria from a variety of organs, beginning with invertebrates and ending with mammals.[11] His most successful explantation was from the testes, ovaries, pancreas, and some other organs. The experiments with liver cells were unsuccessful.

The lower the position of an organism on the family tree of the organic world, the more primitive it is, the more successful is mitochondrial explantation.

Infection of an organism by bacteria from the cultures produced no harm.

In their experiments, Portier and his followers used all the achievements of the newest bacteriological methods. The bacteria that they cultured are identical with true mitochondria. Portier gives many details on experimental methods, and features of bacteria-mitochondria. The reader is referred to his book (Portier 1918).

In the opinion of Portier and others, mitochondria are able to leave their cells and lead free, autonomous lives, as "blood plates," where their independence and bacterial nature are exhibited more clearly.

Portier claims that organisms are in constant need of replenishment of their symbiotic bacteria-mitochondria, and that these are delivered from the outside. In other words, [life forms need a] constant infection (to be compared to infection by zoochlorellae and zooxanthellae).[12]

Bacteria-mitochondria are delivered into the animal organism especially with fats, and into the plant organism with fungal mycorrhiza (see below) and other such structures, Portier demonstrates.

Food is therefore not just delivery of nutrients but also delivery of live organisms, symbionts that are structural parts of the cells of the organism that consumes this food.

Based on many experiments, Portier and other French researchers proved that avitaminosis (starvation of an organism by lack of vitamins) is nothing else but asymbiosis—the lack of symbiotic bacteria. When bacterial cultures were introduced into rats starved by avitaminosis, they rapidly lost the symptoms of starvation; under repeated infection, animals regained health (Portier 1918).

Those who accepted the bacterial nature of mitochondria (Portier [1918], Meves [1918], Wallin [1922–1923]) met with sharp opposition from a number of representatives of orthodox science (Lumière [1919], Buchner [1921], etc.).

However, the objections by themselves were not convincing, especially because they were clearly caused by controversial treatment of mitochondria, as explained above, compared to the established tradition. The objections consisted only in unmotivated suspicions that experiments of Portier and others were methodologically flawed! High expectations have been placed in the Janus green B dye, to which, allegedly, bacteria and mitochondria react differently (Cowdry and Olitsky 1922), but they have not been realized.

The bacterial nature of mitochondria, claimed Wallin (1922–1923), who attempted to test Portier's hypothesis in the pages of serious journals, follows from their similarity in shape, reactions with staining dyes and chemical reagents, and physical properties. Wallin states that Portier's views "are in complete harmony with many biological facts."

Indeed, Buchner [1921] himself finally admits that cell organelles currently considered to be mitochondria could de facto prove to be true bacteria, and that "blood plates" may represent symbiotic cytodes that live in blood. Portier's opponents already concede serious points.

I discuss other cell organelles with most intimate connection to mitochondria; they may represent mitochondrial modifications. If shown to be true these other organelles are of the same nature as mitochondria. They are symbiotic cytodes or entire systems of such organisms.

5. Ergastoplasm [Endoplasmic Reticulum]

In energetically active cells of glandular epithelium, one can see irregular convoluted and globule-forming filaments, bands, and plates, which are stained in the same fashion as nuclear chromatin. These are so-called ergastoplasmic structures or ergastoplasm[13] [endoplasmic reticulum]. When the gland cell activity more or less ceases, these structures can disappear.

New studies show that "the major part of ergastoplasmic structures is identical to mitochondria" (Maksimov [1914]).

6. Golgi Apparatus

A loose network or grid around the nucleus, consisting of densely staining convoluted fibers. This apparatus was first described from nerve cells, and then found in other animal cells. Something similar is known in plants as well (e.g., in roots of grasses.)

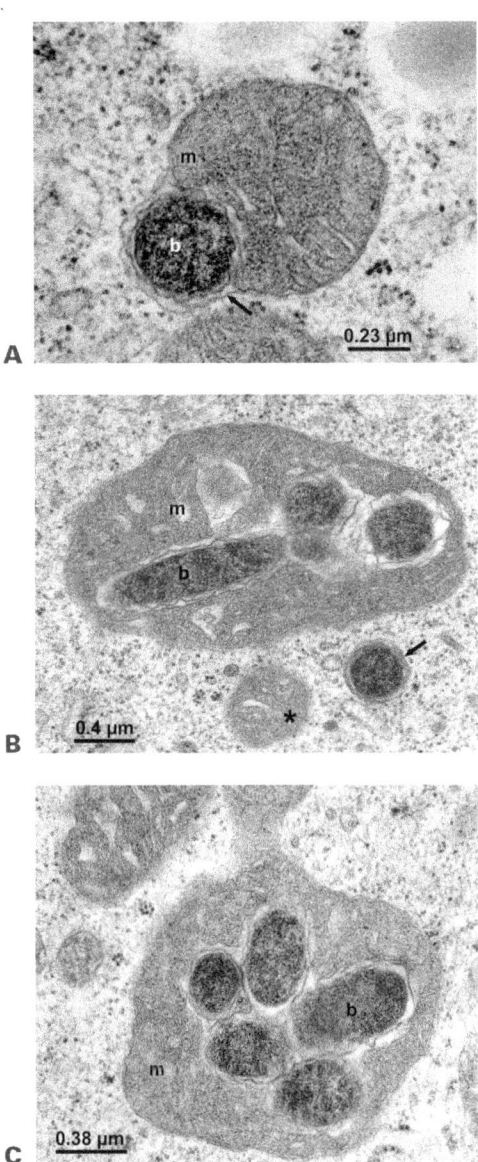

Figure II-4. Bacterial symbionts *(Midichloria mitochondrii)* inside mitochondria of ticks. Kozo-Polyansky would be delighted to know about recently discovered endosymbiotic bacteria that live *inside* endosymbiotic organelles of animals! A, Previtellogenic egg of *Ixodes ricinus* tick with *Midichloria* (b) inside the outer (arrow) mitochondrial membrane; m = mitochondrion. B, A mitochondrion (m) infected by three *Midichloria* (b). Arrow = a single *Midichloria* free in the egg cytoplasm; asterisk = mitochondrion. C, Five *Midichloria* (b) enclosed within the matrix of a mitochondrion (m). TEM micrographs. (Courtesy Luciano Sacchi).

This apparatus is inherited during cell division. Along with karyokinesis, one can observe "dictyokinesis"—division of the reticular apparatus into two parts.

"There are indications that at least in some cases the 'Golgi apparatus' co-incides with the mitochondrial apparatus of the cell" (Maksimov [1914]) or represents its artifact, i.e., a result due to the treatment (Drew 1920). On the other hand, this apparatus highly resembles a column of symbiotic bacteria that surround the nucleus in *Pelomyxa* amoebae (see above, chap. I.4). Those bacteria could easily be mistaken for a "Golgi apparatus."

Some authors even think that the "Golgi apparatus" is identical to chon-driome or "ergastoplasm." On the contrary, Guillermond [and Mangenot] [1922] and others attempted to compare "Golgi apparatus" to "Holmgren's system of canals." Such "canals" (which in fact may be not canals at all) stretch in some kinds of cells from the surface toward the nucleus. Such structures could represent combinations of vacuoles (Guillermond [and Mangenot] [1922]).

7. Nerve Fibrils of Němec

In plant root cells the longitudinal bundles of fibrils are directed upward and serve to conduct stimuli, claims Němec [1901, 1910] who discovered them a long time ago.

According to the new studies by Bambacioni [1920], these fibrils in some cases are long and convoluted mitochondria, in other cases elongated vacuoles.

8. Physodes [of Brown Algae]

These plant organelles,[14] discovered by Crato (1893) in brown algae, are ve-sicular bodies, which consist of phenol-like substances. They consistently contain phloroglucinol. They are located between granules and vacuoles in cytoplasm and characteristically move on their own.

Some may be equivalent to vacuoles such as fucosane vesicles (Linsbauer [1917]), but in general they give an impression of independently living units, or, as Crato originally thought, they are a special kind of mitochondria, i.e., in our understanding, cytodes.

9. Myofibrils (Contractile Fibers)

These organelles, characteristic of animals, resemble mitochondria (see also pp. 50–52 on cytoplasm structure), and on the other hand, also cytodes.

In special cells of some ticks, among other symbiotic microorganisms, Reichenow (1922) discovered a remarkable kind of "bacterium." "It consists of a thin (about 0.5 μm), very long filament, which under staining (by Delafield's hematoxylin) shows the alternating light and dark portions." This filament "imitates a striated muscle fiber." The bacterial nature of this organism was proved.

"Currently, it is firmly established that chondriosomes [mitochondria] take active part in myofibril formation; myofibrils can be viewed as specially differentiated chondriochonts" (Maksimov [1914]).

Therefore, through various approaches one can come to an idea that contractile fibers, just like the "organelles" discussed above, also belong to cytodes.

10. Blepharoplasts [Cytoplasmic Bodies That Bear an Undulipodium]

The term "blepharoplast"[15] is used for a cytoplasmic body that bears a flagellum [undulipodium], or several, that extends from the cell. Blepharoplasts are found in flagellates [mastigote protists], in [undulipodiated] cells of sponges, and also in spermatozoa, not only those of animals, but also of plants.

Many consider it proved that blepharoplasts are modifications of centrosomes (or centrioles). The blepharoplasts are able to develop into the centrosomes, and vice versa, depending on the life history stage to which they belong. If for animal cells this is hardly contested (Maksimov [1914]), then it is most likely also for plant cells, judging from sperm formation in cycads (Chamberlain [1921] and others).

Blepharoplasts, in many, or maybe in all, cases, are equivalent to centrosomes or centrioles. Since connection of the centrosomes or centrioles with plastids was established (p. 31), then blepharoplasts are equivalent to plastids. They are, so to say, flagellated plastids—or, in other words, flagellated cytodes.

Since we recognized above, in various cells' organelles, quite variable models of cytodes, we should not be surprised that finally we see here a flagellate [mastigote] model.

Cell division begins with the division [development or replication of the unit centriole] of the blepharoplast in its role as a centrosome [large sized "cell center"] or [unit] centriole [small sized "cell center," Margulis, 1993, chap. 8; Chapman and Alliegro 2007]. The blepharoplast, a motile cell's organelle or partner [limited to sperm-forming cells, motile algae or other swimming protoctist], bears one or more flagella [undulipodia].

Here, exactly the same phenomenon takes place as in a bacterial consortium in the genus *Chlorochromatium* and other unnamed cases, described by Pascher [1914] (see pp. 11–13), where division of the system (I call it a model of the cell) begins with division of the flagellated [undulipodiated] partner.

Similar to the use of the flagellated partner in bacterial consortia for the locomotion of the entire system, the blepharoplast is also used for the locomotion of sperm.

Finally, similar to a variety of shapes of flagellated partners in bacterial symbiotic systems, up to spiral ones (spirilla), blepharoplasts repeat the same forms, up to the helical (Webber [1897, 1901]; Chamberlain [1921]).

Cilia of the cells of ciliated epithelium, etc., also often are associated with special bodies embedded into cytoplasm called "terminal buttons" or "basal bodies" [kinetosomes]. Unlike in blepharoplasts, no connection of basal bodies with centrosomes exists in these cases. Those ciliated kinetosomes, nevertheless, highly resemble cytodes [bacteria] known as *Cyanotheca longipes* (Pascher [1914]), which live in the mucus of spherical colonies of some protozoa and extend from it a long flagellum [undulipodium].

A suspicion of the bacterial nature of those "kinetoplastic" [locomotory] organelles is, without a doubt, quite legitimate.

11. Elaioplasts [Oil Bodies in Monocots]

These are enigmatic "oil bodies," e.g., in monocot plants, that consist of cytoplasmic stroma and multiple tiny oil drops. These organelles[16] originate through the fusion of plastids [chloroplasts] (or mitochondria), according to Beer (1909).

Probably to the same group belong "placoplasts," "sparsioplasts," and "stabiloplasts" of the diatoms and "linoplasts" of Peridinea [dinomastigotes]—organelles, which have not been studied at all and which, of course, are "enigmatic" for traditional science.

12. Aleurone [Proteinaceous Granules of Seeds]

Aleuronic, proteinaceous granules, or, in short, "aleurone,"[17] are characteristic of the seeds of many plants. In addition to their chemical composition, they can be recognized by their appearance; they are dense, shiny corpuscles of spherical or ellipsoid shape. They are considered storage of nutrients, and, genetically, a result of "vacuole compression."

Such granules are typically expressed in grasses, where their origin was studied in detail by Peklo [1913]. Here, they are very small, 0.3–0.5 µm, and fill, in large amount, certain cells with especially dense cytoplasm content. Such "aleuronic" cells are the outermost cells of the endosperm in the grass seed, which form a homogeneous, or more rarely multilayered (e.g., in *Hordeum coeleste*), "aleuronic layer."

This layer can be recognized not only by its content but also by the shape of the cells themselves; they are smaller than true endosperm cells and have a more regular cuboid shape.

A certain regularity exists in distribution of those granules in a cell: the granules are positioned in rows, or beadlike (Pfeffer [1881] and Harz [1885]).

The aleurone layer continues also into the seed lobe, where it generally has the same cell structure.

The most recent studies (Stoward 1911; Grüss 1912) without a doubt proved not only that the aleurone layer produces diastase and other enzymes, but also that the majority of enzymes, which are needed for dissolving and digestion of starch contained in endosperm, are produced by the same layer.

In the seeds of a grass *Lolium temulentum*, outside of the aleurone layer and also between its cells, Vogl (1898), in a wonderful discovery, observed the constant presence of an enigmatic fungus and suggested that the very toxicity of this grass (its diagnostic character) is connected to the presence of this fungus.[18] This fungus is necessary for the existence of *Lolium* (Freeman 1913). Seeds stripped of this fungus do not germinate well, possibly due to a lack of necessary enzymes.

These data caused Peklo [1913] to question whether formation of the aleurone layer in grasses in general is connected to the activity of a specific fungus.

Mycorrhizae in the cells of the aleurone layer in various kinds of grasses were discovered by Peklo [1913] in a study based on the best techniques. All aleurone cells have similar structure. Most of the cell is occupied by a globule of hyphae, or mycorrhizal threads. Often, the hypha is thick and with only one coil; more rarely it is thin and coils many times. This fungal globule, when viewed superficially, gives an impression of "dense cytoplasm."

Fungal hyphae lack walls between cells, so the fungus is likely unicellular. Nuclei and vacuoles are inside, but the cell wall cannot be seen.

The entire surface of the hyphae is covered by shiny, wartlike protuberances. These protuberances, due to their optical features, are clearly visible, while the rest of the fungus remains indiscernible without prior treatment.

Those shiny protuberances of fungal hyphae are aleurone "grains." Their development can be traced, beginning from a tiny tubercle on a hypha, and until complete formation. Well-developed "grains" retain their connection with hyphae.

Aleuronic cells of scutellum are the same. The same fungus is inherited, at least from the female side, although its transmission with the pollen granules is also possible.*

Possibly, the enzymatic features characteristic for the aleurone layer in fact belong to the symbiotic fungus.

Among free-living fungi, a mold, *Amylomyces rouxii*, most resembles the aleuronic fungus, according to Peklo [1913]. It is known precisely for its ability to utilize starch. This fungus also has extensions on the surface of its hyphae, equivalent to aleuronic granules of the fungus symbiotic with grasses.[†]

Aleuronic granules in other plants, e.g., in legumes and castor bean, Peklo [1913] thinks, may have the same nature as these "grains" in grasses.

13. Cytoplasm [Liquid Homogeneous Substance That Contains Once-Autonomous Organisms]

Therefore, some of the cell's organelles undoubtedly represent once-autonomous organisms, or even combinations of such organisms. For other organelles,

---

*Guilliermond found grains also immediately next to the growth point.

[†]Meyer [1923] "criticizes" Peklo [1913], stating without any proof that this author dealt not with "real" aleuronic grains. Although unfounded comments may be ignored, we need to note, however, that even a nonexpert recognizes aleuronic grains; and on the other hand, if Peklo dealt not with real grains but just fungal extensions, the latter ones are remarkably similar to the grains, which is also intriguing.

we do not have definite data, but we are entitled to the opinion that they are similar to the former, from which we conclude that they are not a result of cell differentiation, but autonomous partners in the system that we call a cell.

All these "structural inclusions of a cell" are located inside the cytoplasm. The nature of this background medium of a cell is completely unknown.

Many researchers accept that cytoplasm has a special organization that cannot be distinguished under a microscope. This organization, however, is depicted differently by different scientists [Derschau 1920].

The cytoplasm has reticulate or spongelike structure and consists of a denser, feltlike substance and another, more fluid one that fills the gaps (reticular substance, enchyleme), according to Frommann [1880].

The basis of the cytoplasm is a liquid homogenous substance, in which dense filaments of "filiar" substance are dispersed, according to Flemming [1882]. These filaments may be grains, form globules, etc. Connected to each other, such filaments create "reticulate" structure.

Cytoplasm has foamlike or alveolar structure, according to Bütschli [1892]. The more liquid substance (enchyleme) fills the alveoli.

Cytoplasm consists of a mass of grains, "granules," which are spherical, rodlike, filamentous, etc. bodies that often form beadlike structures, according to Altmann [1890]. The substance between these granules is the "passive part of a cell."

Granules or "spherules" can greatly vary in their shape and size, and their distribution and combinations are very complex, according to Kunstler [1889]).

Finally, according to Fayod [1891], cytoplasm is built of elements called "spirosparts"—complex structures that are made of long spiral "tubules" or filaments (spirofibrils), wound around each other in a certain way. Crossing spirosparts create an impression of a network with loops and alveoli.

All these authors observed quite real things. Their contradictory descriptions are explained if the cytoplasm in different cells differs in structure, or if the same elements undergo structural transformations, or if the same appearance permits alternative interpretation.

Bütschli's alveoli may be equivalent to Altmann's granules and Kunstler's spherules. Flemming's filaments may be weaved into Frommann's network, and under a close observation sometimes show the features of Fayod's "system of tubules."

The theory of Fayod [1891] was considered "improbable" by Borodin [1910], but recently it was confirmed by G. Entz [1881] and his student Francé [1908].

The structure of "cytoplasmic elements" is even more complicated and diverse than it was thought by Fayod (Entz [1881]).

What are these cytoplasmic structures?

Altmann [1890] concluded that cytoplasm is a kind of zoogloea; its "granules" are autonomous bioblasts. According to Altmann, cytoplasm is a collection of cytodes—"the most elementary organisms," i.e. a symbiotic system.

Independently, studies of Fayod [1891] led to the same conclusion. "It is remarkable,—he wrote—that among pathogenic and many other bacteria we clearly find all forms that correspond to various cytoplasmic elements" and their combination. Some types of cytoplasm, according to Fayod, bear an incredibly close resemblance to the colonies of colorless blue-greens such as *Gloeocapsa, Nostoc,* etc.

If cytoplasm consists of elements various authors assign to it, it resembles a concentration of cytodes. The majority of authors tend to see those granules, filaments, tubules, etc., not as elements of cytoplasm themselves but as organelles such as mitochondria and their modification, e.g., Golgi apparatus, ergastic formations [endoplasmic reticulum].

Regarding granules, filaments, and spirosparts as cytoplasmic inclusions, Meyer [1920], Lundegardh [1922], Tischler [1917], and other most recent authors think that cytoplasm itself is an "amicroscopic emulsioid water solution." The chemical nature of this solution remains unknown.

We should not forget that "true protein substances are found not so much in the cytoplasm itself but rather in some cytoplasmic structures" (Borodin [1910]), i.e., in those same mitochondria, filaments, granules, networks, etc., and that it is those protein substances that are considered characteristic of cytoplasm.

Regarding the structure of such a cytoplasm, Meyer [1923] suggests a rather complicated and absolutely fantastic "theory," which I cannot discuss here.*

---

*Cytoplasm consists of numerous "smallest hereditary machines" that have "an incredibly complicated structure" and "are not made of molecules and electrons" but of "mions," the units two thousand times smaller than an electron. These "machines" are called "vitules." A "vitule" is an element of a cell from which all organelles and other inclusions are removed.

Even if so, we must agree that a cell represents a collection of heterogeneous, autonomous units of life—"biotes,"* which lead a symbiotic way of life embedded in a problematic "cytoplasm" like cytodes in zoogloeas in their "jelly" that makes their foundation and cements them together.

*The term "biotes" ["i bioti"] is proposed by Bolaffio (1922) to indicate noncellular organisms, which together make up a cell.

III

Multicellular Organisms

A. First Series of Examples

1. Lichens

Lichens[1] are thallophytes. Their thallus may be either flat and crustose, or fruticose. In a crustose thallus, there is usually the following set of multicellular layers: the upper, thick, often brownish layer; then a colorless layer; and, finally, a thick layer with rhizoids that serve as roots. Cylindrical thalli have a peripheral thick layer, then a green layer, and, in the center, a colorless core layer. The core layer also includes longitudinal mechanical fibers.

These layers—which they called a bark, or cover layer; a green, or photosynthetic, layer; and a core, or mechanical, layer—were already distinguished by earlier authors. They were considered true tissues, which are a product of anatomical differentiation. Green cells, for example, were considered a modification of colorless cells. Nobody had any reason to doubt that lichens are plants that deserve a separate taxon systematically equivalent to such taxa as mosses, algae, and fungi. Due to the presence of chlorophyll, in the early nineteenth century lichens were considered to be close to algae. Later, it was found that they have "pouches" (asci), a feature that places them close to fungi.

Closer study of lichens revealed that their colorless parts are strikingly similar to fungal tissue, while the elements that composed the green layer resemble green unicellular algae. A double assumption based on these facts was made by de Bary, in the 1860s. Either (1) certain unicellular algae repre-

Figure III-1. Lichen *(Cladonia cristatella):* A protoctist (algal)-fungal symbiosis.

sent the non-photosynthetic elements of lichen green tissue, or (2) the lichens are simply fungi that incorporate [green] growing algae.

Soon the second assumption was experimentally proved to be correct. The green elements do not originate from, and are not transformed into, the colorless ones. It was the luck of Famintsyn and Baranetsky [1867] [and Baranetsky (1868)] to be the first to isolate and culture separately a number of green unicellular algae from lichens. These algae were what the elements of green tissue proved to be. Some common, free-living algae were identified as lichen components. Blue-green algae from green tissue of other lichens, where they formed "green" tissue, were isolated by Itzigsohn [1868].

The partner fungi, as opposed to algae, were hard to culture separately. With a few exceptions, they are not found in free-living condition in nature.

Some time after the successful separation of lichens into their fungal and algal components, an artificial synthesis of quite viable lichens was achieved from those algae and fungi that earlier lived separately and independently.

One should not think that lichens are just a simple sum of certain algae and fungi. Rather, they have many specific features found neither in algae nor in fungi. The following features belong to the complex lichen body but not to its separate components.

The structure of the thallus, characteristic for any typical lichen, found neither in the algal nor the fungal partner, is produced by their interaction. This interaction can be traced.

Algae flourish and reproduce in many cases depending on the conditions favorable for photosynthesis. Therefore, their distribution depends on light. Location of its algae determines the entire architecture of the lichen. Transition from crustose to fruticose forms, i.e., from flattened to vertical development, according to Elenkin and others, is determined by dependence of their "green tissue" on light.

The amazing hardiness of lichens is in contrast to their components taken separately. Lichen prospers in conditions not tolerated separately by either algae or fungi. Both partners appear healthier when they participate in lichen formation than they do solo, outside of it,

Lichens have characteristic modes of reproduction. Algae that reside in the thallus stop reproduction by zoospores, i.e., they lose typical algal motile propagules and reproduce by simple division. Fungi develop their own pouches [asci], and, in very rare cases, basidia. When fungal spores are released, cells from the green layer are sometimes also released. The fungal spore germinates near its alga; i.e., conditions for symbiosis are present from the very first step.

An especially typical reproductive device in lichens is the so-called soredia. These are multicellular extensions of the thallus that consist of several green elements, i.e., algae, entangled by colorless threads of the fungus. Such soredia are often concentrated in numbers within special receptacles, called "soralia"— sort of "fruiting bodies." These accumulations result from the metamorphosis of fungal fruiting bodies penetrated by algal colonies (Reinke [1901]).

We thus witness an increasing substitution of the original reproductive mode of the fungi and algae, of the lichen consortium, by a reproductive mode specific for the united body.

The products of lichen metabolism are: lichenin, a substance similar to starch; numerous, about 200, different "lichenic acids"; and specific lichen pigments. These substances are unknown in lichen-component partners studied individually. Evidently these metabolic products result from the composite organism.

Thus, everywhere—in its chemistry, its shape, its structure, its life, its distribution—the composite lichen exhibits new features not characteristic of its separated components.

Modern science cannot yet tell what exactly are the relationships of fungi and algae in lichens, what physiological cement makes these partners whole. Opinions differ. Here are the most important ones:

1. Fungi and algae are mutually useful to each other (mutualism): in lichens, a "physiological collaboration" (of course, unconscious) takes place. According to Reinke, "green elements, so-called gonidia, like green leaves, serve the composite organism through manufacturing of organic substances; mycelium serves as both stem and root, delivering water along with salts taken from the soil." Another opinion (Tobler [1911]) is that the alga utilizes carbon from the products of fungal metabolism, while the fungus takes organic substances from the alga. Another feature is that the alga uses the fungus as a comfortable shelter, as protection from desiccation.
2. One side exploits the other.
    (a) Fungus is an "endosaprophyte," in the opinion of Elenkin [1921], and feeds on dying algae. This opinion is based mainly on the fact that the thallus contains dying, "necrotic" algal areas that are consumed by the fungus.
    (b) Fungus is a parasite of alga. This opinion is based on the fact that the fungal threads possess special extensions that penetrate the green cells.
    (c) The relationship of fungus and alga is the same as in humans and cattle: the fungus feeds on algal offspring (helotism, in the opinion of Warming [1901]).

With the exception of the helotism hypothesis (which assumes in a fungus the existence of free will and conscience, and even rationality), unfounded and anthropocentric, all others are supported by data.

One, however, should not forget that the green coloration of the algae is far from a universal proof of their independent feeding. Chlorophyll sometimes serves not for photosynthesis, but to enhance the processing of already existing organic food.

An example of parasitism of a *green* alga on a colorless fungus was discovered by Kohl [1903] and others.

Lichens often inhabit places where the light conditions make photosynthesis barely possible. In other cases, the bark layer of lichen thallus is impermeable to both air and sunlight, and therefore the possibility of photosynthetic activity in green tissue can be considered practically impossible.

Lichens highly depend on the soil on which they develop. The fungus in their thallus remains either saprophytic or parasitic, depending on the substrate on which the thallus grows. In feeding, the fungus may be rather independent from the alga.

Without discussion of the complicated issue of the physiological relationship between lichen partners, we admit that whatever its nature, lichens are indeed composite organisms. This admission suffices for now.

We say, following [A. L.] Smith (1921), that "development of a lichen is the result of a mutualistic interaction of two organisms; yet, since the fungus creates the reproductive organs, it is the dominant partner. Algae take a subordinate position, although they are important in determination of the shape and structure of the thallus." "Simple parasitism cannot explain this association of the two plants, when the removal of one symbiont—no matter which, fungus or alga—ends the lichen's existence. Together they form a stable unit that develops, varies, and progresses in new directions. The constant characters in lichens are transmitted here in the same manner as in other *units* of the organic world" [Reinke 1901].

2. Plants [Successful Grafts]

Synthesis of flowering plant types into different composite organisms is practiced in horticulture, via the well-known method of grafting.[2]

When common grafting is performed, some notable observations may be made.

Grafting is successful between the races of the same species, but also between quite different species, and even between different genera. "The limits of vegetative combination of different plants are much broader than for sexual combination in hybridization. Thus, representatives of different genera of many families (e.g., Compositae, Rosaceae, Solanaceae, Cactaceae) are easily transplanted (grafted) to, and grow well on, each other, while sexual hybrids between genera are rare exceptions." "In cacti, it is possible to graft a morphologically quite different, thin-stemmed *Pereskia* on the globose *Echinocactus*, and vice versa. It is also possible to successfully transplant such morphologically different forms as herbaceous plants to woody plants, e.g., species of *Abutilon* on *Modiola*" (Vavilov [1916]).

Grafting, in some taxa, between more distant species has been more successful than between closely related ones. Most pear varieties are better grafted to quince than to apple. Potato, which belongs to the nightshade genus, *Solanum,* is more easily combined with jimson weed *[Datura]* than with other nightshades.

Ten or a hundred different scions can be successfully grafted to a single stock. Often, to make "three-tier" grafting is practical: a desired plant is grafted to a stock, which was earlier grafted on a third plant (basic stock).

Mutual influence of the partners in the grafting symbiosis is expressed only in their nutrition in all these combinations. These symbionts, beginning from the very point of their connection, retain complete individualities in their diagnostic characteristics. "Dozens and hundreds of apple varieties, grafted sometimes just out of curiosity to a single tree, and retaining at the same time their specific form, represent an obvious example of a graft's lack of influence on the stock" (Vavilov [1916]).

Remarkable phenomena are sometimes exhibited in plant nutrition. The legume *Cytisus hirsutus* (hairy broom), for example, develops especially well on the roots of golden rain *(C. laburnum),* even "much better than on its own roots" (Baur [1922]).

Different organism combinations as scion and stock in grafting are called chimeras. However, in botany the term "chimera" is limited to certain grafting products, namely, for those that combine both scion and stock characters at the same time (Winckler 1907).

Such combinations can be of two types.

The first type of chimera is illustrated in *Pelargonium.* In grafting between white-leafed and green-leafed races, shoots that emerge along the line of connection carry both green and white leaves as well as half-colored: one (longitudinal) half green, another white. Another such example is presented by combinations of tomato and black nightshade—the two species differ in leaf segment shape as well as coloration, shape of corollae, and other characters. Leaves and flowers of this chimera are divided into sections—some show characters of tomato, others retain those of black nightshade.

The second type of chimera does not produce such a mosaic picture. They have features intermediate between those of scion and stock and thus appear to be true hybrids, hybrids produced vegetatively [that bypass sex].

Especially famous are the following examples. Bud eyes of a small shrub, red-flowered *Cytisus* [now *Chamaecytisus*] *purpureus* were grafted to *Laburnum vulgare* [now *L. anagyroides*], a small tree with yellow flowers. Shoots appeared along the connection line that could be interpreted as a hybrid, in 1829. This chimera was reproduced by cuttings and became widespread in horticulture under the name *Cytisus* [now *Laburnocytisus*] *adamii,* named after the horticulturist [Jean Louis] Adam, who grew it. These plant species were not successfully hybridized by crosses.

A chimera may form shoots, leaves, and other features that completely exhibit characters of one of the parents.

Similar examples are: *Crataegomespilus* (*C. dardarii* and *C. asnieresii*), a "hybrid" produced by grafting hawthorn *(Crataegus monogyna)* and medlar [*Mespilus*]. A number of nightshade species (*Solanum tubingense, S. koelreuterianum, S. gaertnerianum,* etc.) were produced from tomato and black nightshade. Also a poplar was synthesized from black poplar [*Populus nigra*] and *Populus trichocarpa.* Similar chimeras are also known in *Pelargonium.*

At first glance one thinks chimeras combine scion and stock elements in a single shoot, a single leaf, a single flower. But chimeras have a propensity for segregation. They form branches, leaves, or flowers whose features are of only one of the parents.

Science has a direct way to recognize "parental" components in the chimeras. Characters such as cell shape, content coloration, chromosome number, etc., distinguish scion cells from stock cells.

Thus, one may follow the formation of a chimeric shoot from its "parental" components from the start, and afterward locate the partners.

"Formation of chimeras is determined by the fact that entire groups of cells belonging to both plant species participate in formation of accessory buds in the area of scion and stock contact." Further, "tissues formed by the cells of two independent plant species, combining in unusual ways, still grow in concert, and produce a shoot composed of tissues belonging to both plants."

In chimeras of the first kind (which are called "sectorial"), one symbiont forms a longitudinal thread (a sector), or a band in the stem, which otherwise belongs to the second symbiont. Organs that grow from this thread have features of this symbiont, while those that appear on the rest of the stem have features of the second. Finally, organs in the contact zone may be asymmetric [resemble one or another of the "parents"] to a different degree.

Chimeras of the second kind, called "periclinal," are formed in a completely different manner. One symbiont in this case appears wrapped around the other, "like a glove on a hand" (comparison of Baur [1922]).

From an anatomical viewpoint, *Cytisus adami* is a *Laburnum vulgare* (a "hand") covered as by an epidermis by *Cytisus purpureus* (a "glove"). *Solanum tubingense* is black nightshade gloved in a tomato. *Crataegomespilus asnieresii* is hawthorn in a sheath of medlar, etc.

A green *Pelargonium* inside the "skin" of a white one was grown by Baur [1922]; a white one clothed in two layers of a green one; a white one in a single-layered "glove" made of a green one, etc.

Thus, one symbiont may enclose another, grow around it like epidermal tissue, and thus appear like only one of the tissues of the composite, chimeric organisms.

Yet symbionts in no way lose their autonomy, their individuality. This we know because parentals may be released from the synthetic chimeras.

"The fact," writes Vavilov [1916], "that these released plants represent forms that are absolutely identical to the original species, notwithstanding the fact that, e.g., in the case of *Cytisus adami,* the epidermis of *C. purpureus* (I would say "*C. purpureus* as epidermis"—B. K.-P.) covered inner tissues of *Laburnum vulgare* for almost 100 years, is the best proof of the fact that changes during grafting do not influence species-specific characters."

However, the fact that purple broom that existed for almost a hundred years as a tissue of Adam's broom was able to restore itself to its former shape does not imply that such restoration of an individual that became a tissue still will be possible in ten thousand or a million years. Most likely, the tissue of such an organism will remain irreversibly composite.

Fusion of growing tissues occurs in these cases, but not fusion of cells or nuclei. Therefore, while "periclinal" chimeras such as Adam's broom have a remarkable resemblance to true hybrids, many researchers refrain from applying the term "hybrid." The formation of true hybrids is contingent on fusion of both cytoplasm and nuclei of at least two cells.

In the "Darwin's nightshade" chimera created by Winckler [1907], the subepidermal layer was produced by fusion of a tomato cell with a black nightshade cell. This cell, in Winckler's opinion, is a true hybrid, or "burdon," though it was created vegetatively [without sex]. Winckler based this opinion on the following. Black nightshade has 72 (or 36) chromosomes in its nuclei, while tomato has 24 (12), and Darwin's nightshade in subepidermal cells has 48 (24). Since, according to Baur [1922], an alternative explanation for chromosomal numbers is possible in this case, Winckler's "burdon" remains in doubt.

Something like chimeras, and even burdons, is known in fungi. In the mold fungus *Phycomyces nitens*, Burgeff [1912] created a mixture that contained nuclei and cytoplasms from two different races of this organism. The resulting fungus seemed to be a "compromise," which later exhibited segregation as in flowering plants.

A remarkably broad *possibility* exists to create such burdons. Naked cytoplasm of rust fungus [one of the "parents"] fused with the cytoplasm of its host grass in such a way that the composite cytoplasm appeared uniform. This mixed condition from two entirely different cytoplasmic sources Eriksson [1910, 1921] called "mycoplasm" [Martin 1922].

Skeptics prevailed. The idea that a cytoplasmic union had occurred between very different organisms, such as a fungus and a flowering plant, was rejected by professionals. However, "this theory [the mycoplasm theory] is one of the most outstanding and brilliant modifications of the idea of symbiosis that has ever been proposed in botany," wrote Whetzel (1918).

3. Animals [Chimeras and Graft Hybrids]

Experimental synthesis of composite units comparable to grafts in horticulture also has been successful in zoology.

Experiments classified grafts as autoplastic (parts of an individual grafted to a different place on its body); homeoplastic (parts grafted to a different individual of the same species); and heteroplastic (parts of one species grafted to a different species).

We mention heteroplastic combinations. Transplantation of separate organs has been especially successful. Many grafted organs were successfully transplanted to a foreign body.

Theoretically more interesting are experiments that combined entire portions of the body and formed composite chimeric entities and "graft hybrids."

Experiments with two hydra species in two different genera, *Pelmatohydra oligactis* and *Hydra vulgaris,* are notable (Issajew 1923). These hydras strongly differ in morphology and coloration.

The chimeras were classified into three types: conplantational, associative, and dissociative (Issajew 1923).

The conplantational type was obtained by longitudinal fusion of two halves of animals of different species. Along the line of fusion, through budding, sectorial chimeras appear that resemble similar combinations in plants. The associative type of chimera is created by fusion of two polyp tubes when one is inserted into another. In "dissociation," individual polyps were cut into very small pieces and mixed. After restoration a mosaic chimera formed.

The following phenomena were noticed in chimeric combinations of entire organisms. All elements of the red hydra *(H. vulgaris)* gradually disappeared and were replaced by the elements of the gray hydra, *P. oligactis.*[3] The sectorial chimera developed into a typical *P. oligactis,* indistinguishable from a normal individual of that species.

However, two sorts of buds were formed when the chimera budded: one was the *P. oligactis* type and the other similar to *H. vulgaris* that later revealed unusual features.

The possibility of "segregation" of "parental" tissue could be explained by retention of some elements of *H. vulgaris.* Intermediate cells of this species do not disappear; rather, they mix with corresponding cells of *P. oligactis* and together form a synthetic cell layer that underlies the ectoderm. This mixing of cells of two species Issajew calls "cytomixis," and corresponding chimeras "cytomictic."

The buds of *P. oligactis,* in a number of generations, produce true hydras of *P. oligactis* species, while buds that resemble those of *H. vulgaris* develop into true graft hybrids, which *combine characters of both species.* This hybrid type, a species created in Issajew's experiments, and nonexistent in nature, he called an "oligactoid."

The further fate of this "oligactoid" is remarkable: it segregates into *P. oligactis* and *P. oligactoides,* in ratio 1:3. One has an impression that a combination of a dominant (oligactoid) and a recessive (oligactis) form takes place, and this is confirmed by observations through a number of consequent generations. Thus, this "oligactoid" is equivalent to Adam's broom.

Butterfly pupae were cut lengthwise, and then Crampton [1898] attempted to fuse halves taken from different species. Although his techniques were quite primitive, fusion usually was successful. Adult butterflies hatched; one half of the body came from one species and the second half from another. Crampton's results were comparable to sectorial chimeras in horticulture.

Similar experiments with frogs, in which young embryonic stages were used, were conducted by Born [1894]. Fusing fragments, Born obtained all kinds of combinations, starting from two-headed monsters and up to chimeric organisms made of combined halves but otherwise normal and reaching adult age.

Similar heteroplastic combination was produced by Harrison [1907]. Following Born's technique, he fused the front portion of a *Rana sylvestris* frog embryo with a posterior portion of an *R. palustris* embryo. What resulted was a viable frog specimen whose dual character was easily seen. Its front portion was dark-colored as in *R. sylvestris,* while its rear, as in *R. palustris,* was light-colored. The boundary between two halves was distinct, although tissue fusion was complete. More than one such specimen was produced.

Dual, composite frogs recall scion and stock: combinations of pear and medlar or American with European grapevines.

Periclinal chimeras have not yet been obtained in animals. Fusions (e.g., like those in Born [1894] and Harrison [1907]) have not been produced in birds and mammals.

We should remember Schmidt's [1920] remark that "if we cannot yet obtain such fusions in higher vertebrates, e.g., in humans, this is exclusively due to the fact that we cannot experiment with their early developmental stages, when such fusions most likely would be quite possible to perform."[4]

4. Consortia of Sponges with Algae

When the first "animal-plant combination" was discovered from Mauritius Island, Areschoug (1854) could not decide whether he dealt with an animal or a plant. He long vacillated before he chose its systematic placement. The presence of green tissue compelled him to assign the enigmatic organism to the algae that he named *Spongocladia.* He noted in its body silica spicules, typical for sponges.

That *Spongocladia* represented a case similar to lichens was first demonstrated by Marchesetti (1884). A combination of a sponge, *Reniera fibulata,* which is also known to live separately, when of course it is colorless, and a filamentous, multicellular green alga, for which the name *Spongocladia* could be retained, was described. Later, other sponge-algae were discovered, in which the same filamentous green alga was present.

A similar double organism is formed by a combination of the alga *Marchesettia spongioides* and various *Reniera* sponge species. This composite is widely distributed, in the Philippines, New Guinea, New Caledonia, Singapore, Madagascar, Celebes, and the Adriatic Sea.

The main body mass in such sponge-alga is formed by a tangle of algal trichomes, and only outside a thin skin of the sponge covers the mass. This provides a model of a periclinal chimera. It is a reasonable question to ask: Who takes charge of the body development and morphology? One cannot help but compare this organism with lichen in which the sponge plays the role of fungus (Buchner 1921).

Another double organism comprises the *Halichondria* sponge and *Struvea*, a green alga. Both partners are found also as separate and independent organisms. The algal body is composed of numerous branches. But when combined with a sponge, the alga forms nonbranching saclike extensions that point in all directions. The sponge, when living independently, forms a flat, crustlike body. In combination with the alga, it develops numerous conical shapes. Sometimes the top of a cone is adorned with a bunch of algal threads.

Even more than *Spongocladia-Reniera*, it is obvious that the *Halichondria* sponge shape, and, of course, its structure, is new relative to each of its separate partners.

Red algae and blue-green cytodes [cyanobacteria] participate in other sponge-algae. The algae apparently are not found without their partners. Sponges that participate in symbiotic union may belong to races that already have lost their ability to live independently.

In dual sponges (*Thamnoclonium*[5] *flabelliforme* and *Reniera fibulata; Scytonema* and *Spongia otochelica*), green, blue-green, or red partners consume the "spicules" and other parts of the sponge's horny skeleton and replace them (Carter [1878]). I compare this to osteoclast activity during bone destruction in vertebrates.

At different depths in the body of the consortium reside different photosynthetic partners. In some *Spongelia*, blue-greens live at its surface, and red algae reside at deeper levels. A diversity of "tissues" results.

Whereas inheritance of the sponge-algae in combination is highly probable, few direct observations have been conducted. In *Spongelia*, for example, planulae larvae that emerge from the sponge's body already contain green symbionts (Schulze [1879]). The symbionts are found already in cleavage stages of the blastula, into which they apparently enter autonomously.

Dual organisms similar to those listed above are widespread among sponges. Brandt [1881] relates this fact to the presence of starch in sponge parenchyma.

The physiological basis of the sponge-algal composite organism, however, is still completely unknown. Algae may be useful to sponges. That some alga-sponge partners are not found free-living attests to the success of symbiosis for green members of this consortium. The constancy of cohabitation, its hereditary character, equally attests to the success of symbiotic association.

B. Second Series of Examples [Plants]

1. Mucous Glands in Aquatic Ferns *(Azolla)* and Hornworts

In a small aquatic "fern" *(Azolla),*[6] each leaf is composed of two lobes, upper and lower, each with a different structure and role. The fleshy upper lobe is a photosynthetic organ [Oes 1913]. Its morphologically lower side faces upward and has photosynthetic tissue. Its morphologically upper side faces downward and possesses a special mucus-secreting cavity. The cavity with its narrow duct opens at the upper side of the leaf. The cavity closes later in life.

The cavities assimilate free nitrogen, N_2. They are the fern's "nitrogen organs."

Colonies of a filamentous blue-green *Anabaena azollae* fill the cavity of this "gland." They [cyanobacterial trichomes] are entangled with threadlike, multicellular extensions of the cavity walls. The plant-cell threads emerge in response to the stimulation by *Anabaena*. In addition, bacteria [and protoctists] may accompany *Anabaena*. Nitrogen is assimilated either by *Anabaena* or by the accompanying bacteria.

This structure of a "mucous gland" [the modified dorsal surface of a leaf] that includes *Anabaena* and bacteria is characteristic for all *Azolla* individuals, regardless of their geographical distribution (America, Asia, Africa, Australia). The gland itself is produced by the presence and action of cytodes [bacteria]. *Anabaena* can be observed in the gland from the beginning of its development. The assumption that the blue-green enters through the duct of the gland is incorrect. [The cyanobacteria do glide into the dorsal leaf surface "glands"; Margulis and Wier, 2005.]

When pondering the plant's "infection," remember that *Anabaena*'s bioblasts [bacterial cells] are found near the plant's growth point [meristem] as well as in closed cavities (indusia) of sporangia.

Similar organs are also present in hornworts.

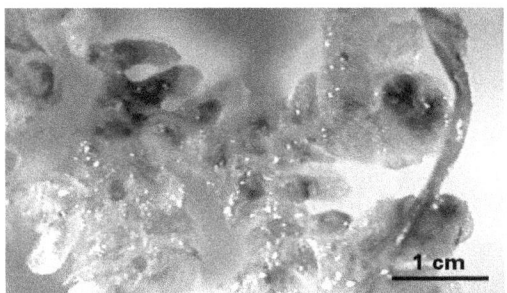

A

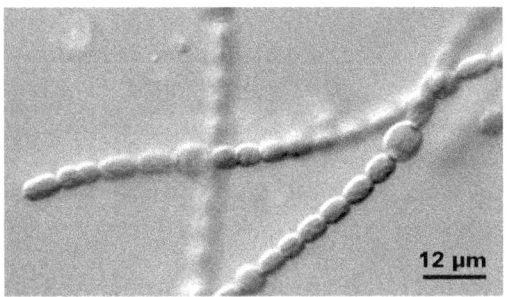

B

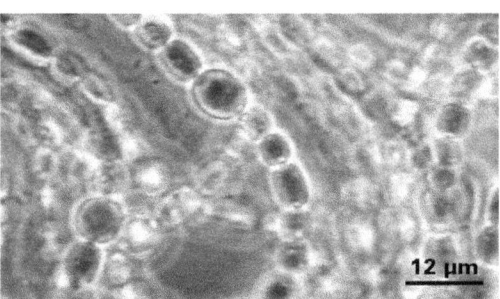

C

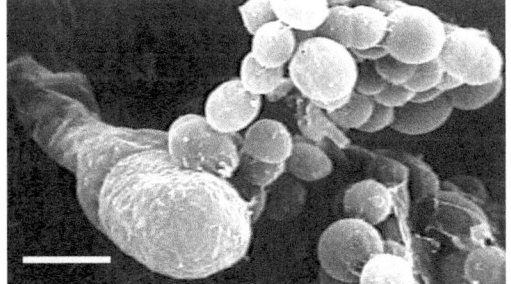

D

Figure III-2. A plant-cyanobacterial symbiosis: *Azolla* sp. (water fern), common in Chinese rice paddies, whose dorsal leaf cavities invariably house the symbiotic cyanobacterium *Anabaena azollae*. A, With the unaided eye and low-power light microscopy the dark cyanobacteria-filled cavities in each leaf can be seen (bar = 1 cm). B, *Anabaena*, the filamentous cyanobacterium grown by itself in sunlight-illuminated mineral nutrient culture displays swollen cells (heterocysts) where atmospheric N_2 is converted into organic nitrogen; differential interference (Nomarski) light microscopy (bar = 12 μm). C, as in B but a standard phase contrast micrograph (bar = 12 μm). D, The protrusion on the lower left in this SEM of the *Azolla* leaf cavity is a specialized plant cell over which the cyanobacterial filaments (the smaller, more spherical cells on the right) are draped. SEM micrograph (bar = 10 μm), courtesy of M. J. Chapman.

The thallus of *Anthoceros laevis,* for example, has characteristic mucous cavities in which dwells *Nostoc,* a blue-green. Moving filaments of *Nostoc* penetrate through the slits of the mucous cavity and produce entire colonies inside. They enlarge the cavity. Basal cells (according to Leitgeb [1878]) divide and form a lid that closes the cavity from outside. Cells of the cavity walls grow into branching multicellular extensions. Those extensions intimately interact with *Nostoc* colonies such that tissue similar to homogeneous parenchyma forms. *Nostoc* is located among the plant cells.

Such glands do not develop when other cyanobacteria, e.g., *Oscillatoria,* enter gland slits.

In other hornworts, e.g., *Dendroceros, Notothylas,* the glands extend from the thallus surface like warts; they are not embedded.

In *Blasia* [a liverwort], colonies of *Nostoc* [the hormogonia of which penetrate the plant tissue] gradually inhabit the "leaf ears" [the periphery of the thallus] due to gliding entry of [the nitrogen-fixing cyanobacterium *Nostoc*]. They increase in size and grow internal [peptidoglycan] wall extensions. These provide an intimate connection of the symbionts [apposed membranes and walls of cyanobacterium and fungus that enhance nutrient transport]. If *Nostoc* is absent, the "ears" degrade.

The fact that in *Azolla,* in hornworts, and also in the flowering plant *Gunnera* (see below), blue-greens and other cytodes [prokaryotes] enter mucous cavities makes one think that they are attracted to those sites. They find shelter. *Nostoc* colonies, due to their water-rich mucus, may also serve for water storage (Goebel [1918]).

2. Stem Glands of *Gunnera*

In all species of the genus *Gunnera,*[7] an angiosperm assumed to be related to *Hippuris,* from South Africa, New Zealand, and South America, one can see, even with the unaided eye, special glands at the periphery of the stem and rhizome, which look like greenish spots [Reimnitz 1909]

These "stem glands," which begin as colorless, are formed in the meristem of the internal growth cone, between each couple of leaves. First one emerges, then the second and the third glands. Each one is divided by mucous canals into several lobes. The mature gland is covered only by a single-cell epidermal layer.

As the gland becomes active, the apical cells of its lobes start producing mucus, and the pressure of the mucus breaks the epidermis. Ruptured glands

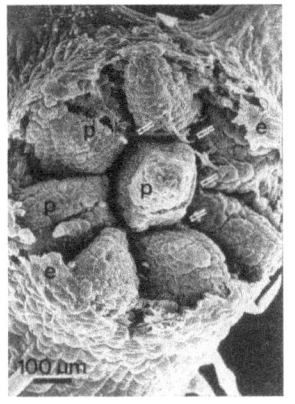

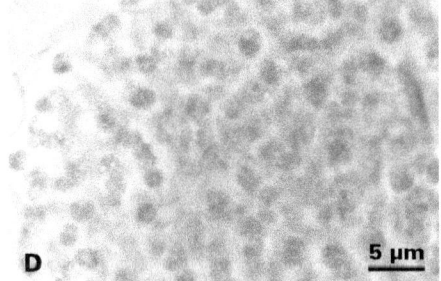

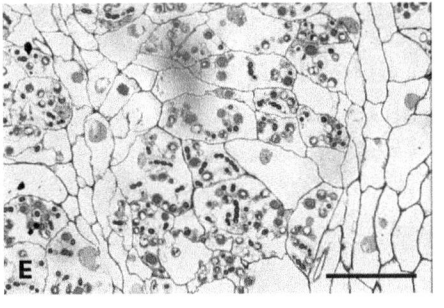

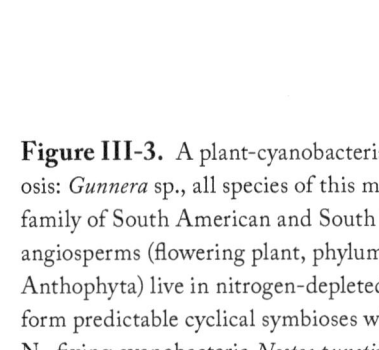

Figure III-3. A plant-cyanobacterial symbiosis: *Gunnera* sp., all species of this monogeneric family of South American and South Pacific angiosperms (flowering plant, phylum Anthophyta) live in nitrogen-depleted soil and form predictable cyclical symbioses with N_2-fixing cyanobacteria *Nostoc punctiforme*. A, B, The Ecuadorian *Gunnera manicata* (called "Poor man's umbrella" or "parasol de los pobres" plant in fields and greenhouses.) With its few huge leaves, *G. manicata* is one of the largest species; it has specialized "bracts," red, slime tissue modified leaves related to transport of its bacterial symbionts from the soil to the glands of each seedling in the inflorescence that bears 150,000 flowers (F); C, seedling with paired glands that are penetrated by and harbor *Nostoc* symbionts. The SEM shows the plant cells that surround the opening into which the cyanobacterial filaments glide; D and E, *Nostoc* in *Gunnera* cells, stained with Giemsa. The heterocystous *Nostoc* has replaced the contents of the plant cells, although their cellulosic cell walls are still intact. Light microscopy, (bars = D, 5 μm; E, 15 μm).

are quickly colonized by blue-greens of the genus *Nostoc: N. punctiforme* (Hariot [1892]), found free-living [in soil] as well, or a specific *N. gunnerae* (Reinke [1873]). Apparently, a certain form of chemotaxis takes place.

Nostoc filaments penetrate into mucous canals; they inhabit the cavities, which are formed by mucus production. From there, the filaments of the *Nostoc* colony penetrate to the intercellular spaces of the surrounding, starch-rich bark parenchyma and closely entangle its walls.

The blue-greens penetrate into cells; they reside in the cytoplasm and mimic chloroplasts. They consume starch and grow to fill the cells almost completely. *Nostoc* appears to be very healthy and disperses and colonizes the bark parenchyma in patches. Cells that contain the blue-green symbionts increase in size but exhibit "no sign of disease" (Tubeuf [1897]).

Mucous canals close by this time, and *Nostoc* colonies become locked inside.

Nostoc, the *Gunnera* symbiont, also lives free in the same localities (Jönsson [1894]). This still needs confirmation.

The physiological side of this cohabitation, and its benefit for *Gunnera*, is as yet unknown. However, the predictable presence of "green glands" in all individuals of all *Gunnera* species in so many countries remote from one another leads us to claim that the symbiosis is neither accidental nor beneficial only for *Nostoc*.

3. Leaf Glands of Plants

Leaf glands[8] are characteristic of many angiosperms, especially from the families Myrsinaceae and Rubiaceae. *Ardisia crispa* from East India has elliptical, simple evergreen leaves with serrated edges. The serrations are inflated and nodular, and by their yellowish coloration contrast with the dark-green background. The nodules are protein glands with complex structure. Their core is occupied by a chamber filled with highly modified cells with a well-developed mass of mitochondria. Walls of the chamber are composed of many cell rows, and the chamber itself is lined by apparent special "epithelium."

Pavetta angustifolia from Java with oleander-like leaves that have glands scattered all over the leaf surface, usually "with a remarkable regularity," is another example [of morphology generated by symbiotic acquisition] (E. F. Smith [1921]). Glands, embedded in leaf tissue, are observed from outside as yellowish nodules.

That these glands are a result of activity of certain microbes was shown by Miehe [1911, 1914, 1919]. Physiological interaction resembles the well-known nodules on legume roots. Unlike legume nodules, Miehe's symbionts are found inside the leaves. Most importantly, they are inherited from generation to generation through the female. How the bacterial symbionts are transported to reproductive organs is not known. Besides the glands, the microbes disperse across other parts of the plant, including the seed. After these microbes are artificially removed, the *Ardisia* glands do not develop. The plant itself fails to develop normally.

Miehe's results have been confirmed (E. F. Smith [1921] and others).

4. Coralloid Organs of Cycads

The representatives of all extant genera of cycads [relict gymnosperms, Cycadaceae] have organs shaped as repeatedly branching coralloid structures [Spratt 1915]. These organs,[9] modified lateral roots, either lie immediately under the soil surface or extend above the surface. Unlike true roots, they possess negative geotropism. Under light, these generally colorless roots become green. A special green zone that underlies the colorless bark is observed.

These organs were considered to be respiratory roots by Reinke.

The hypothesis that coralloid organs not only serve for aeration but at the same time are organs that fix atmospheric nitrogen was first suggested by Life (1901).

The coralloid organs develop due to the action of the nodular bacterium *Bacillus radicicola*. Usually it is also joined by *Azotobacter* and the blue-green *Anabaena*. The presence of *Anabaena* gives these organs their green color.

The green zone is formed by a system of large intercellular spaces, filled with colonies of *Anabaena* and *Azotobacter*.

Special teat-like cells from the cycad plant cross and penetrate the intercellular zone. The cytoplasm that lines the green zone both inside and outside the plant cells harbor the cytode *Bacillus radicicola*. Therefore this cytode imitates mitochondria.

Bacillus radicicola that enters the plant through root pili [one-celled root hairs] is attracted chemotaxically by [the filamentous nitrogen-fixing cyanobacterium] *Anabaena*. "Infection" of plants by *Anabaena* is achieved by a "lens belt" formed at the base of each inflated structure in the coralloid body.

Figure III-4. A plant-cyanobacterial symbiosis: coralloid roots of cycads, evergreen gymnosperm plants (phylum Cycadophyta); their fossils date to the Permian period.

Three kinds of organisms act symbiotically, all useful for cycad plants because they deliver processed chemicals. These microorganisms benefit from the symbioses as well.

5. Mycorrhiza, Plant-Fungal Roots

The name "mycorrhiza" was given by Frank (1885) to an intimate combination of the roots of plants with fungal mycelium.[10] "Just like a combination (synthesis) of an alga with a fungus creates a new individuality—the lichen, symbiosis of a root (i.e., an organ of higher plants) with a fungus leads to the formation of a new morphological unit—a root fungus, or mycorrhiza" (Neger [1913]).

A typical ("ectotrophic") mycorrhiza "is easily recognized by its appearance. Mycorrhizae differ from ordinary roots by their white color, larger thickness, retarded longitudinal growth and sometimes, e.g. in pines, by dichotomous branching, and always by coralloid appearance and absence of root hairs."

The sheath formed from mycelium either has an almost smooth surface or has fuzzy hyphae tips that emerge like a brush.

The mycelial roots are called "rhizoids" by Noack and others, since they probably replace missing root hairs. Magnus [1906] called them "absorbing hyphae."

The internal fungal sheath tissue is not distinctly separated from the plant's root tissue. Mycorrhizal hyphae not only entangle the external cell layers of the root but also penetrate inside. They are limited in their distribution by intercellular spaces. Magnus calls intercellular mycelium "exchange hyphae," since extensive development of its surface and intimate contact with root tissue provide nutrient exchange for both symbiotic partners.

Mycorrhizae were described from Carboniferous fossils (Weiss [1884).

Numerous observations and experiments demonstrate that mycorrhizae are broadly distributed in various plant groups in all parts of the world. They play an important role in the life of the plants that possess them [Sukhov 1914, Fuller 1922].

Some plants such as beech, many conifers, etc., cannot even exist in otherwise normal conditions if these symbiotic fungi are artificially removed.

The most reasonable hypothesis appears to be that of Stahl [1900]. He suggested that mycorrhizae "procure large amount of very specific salts under low transpiration." A symbiotic fungus helps plants in the struggle against other competing fungi. Mycorrhizal fungi may assimilate free nitrogen and provide possibilities for the existence of microorganisms in nitrogen-free soil.

The nature of mycorrhizal fungi and their manner of synthesis are not yet clear. Earlier authors (Rees, Frank, Noack) thought that mycorrhizae can be formed by rather different fungi, which are common also in independent form, while "infection" happens from the soil. A considerable disagreement exists about which fungi form mycorrhizae.

Recent experimental studies (e.g., Fuchs [1911]), as well as attempts to analyze and synthesize artificial mycorrhizae, show the situation not to be so simple. Isolation of fungi from mycorrhiza was not successful. Infection of sterile seedlings with common humus-living fungi did not give rise to mycorrhizae. The formation of mycorrhiza depends not on "infection" by free fungi resident in soil, but is rather determined by hereditary transmission of the symbiotic fungus with the seeds of the partner plant, according to Fuchs (1911).

The presence of specific "fungal spores," for example, in the seeds of conifers, which at the same time are characterized by typical mycorrhizae, was observed (Fuchs [1911]; Neger [1913]). The most recent studies on heather (see pp. 77–78) render this suggestion highly probable.

The symbiotic fungi probably obtain, to some extent, safe shelter and procure various carbohydrates from the root. Fungi develop especially well as mycorrhizal components, in any case.

6. Orchids: Roots, Tubers, and Flowers

A different picture, both morphologically and physiologically, is presented by the symbiosis of orchids and fungi, a classical example of "endotrophic mycorrhiza."[11] This phenomenon, characteristic for the orchid family, is also known in other plants.

Horticulturists have long realized that orchid seeds do not germinate on their own. They germinate only if their embryos are infected by a certain fungus—if a symbiotic synthesis takes place (N. Bernard [1909]).

Separate cases of successfully grown orchid seedlings without the fungus, done by Bernard himself, and later by Knudson [1922], do not change the rule of the necessity of this fungal symbiosis. In natural conditions orchids do not grow on a highly concentrated nutrient medium as in experiments of

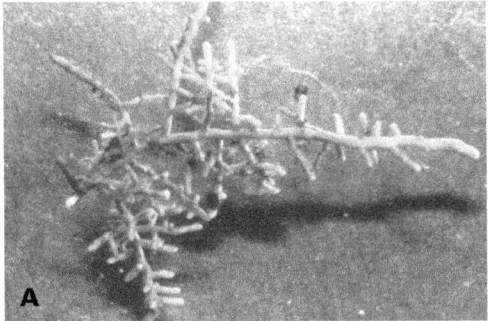

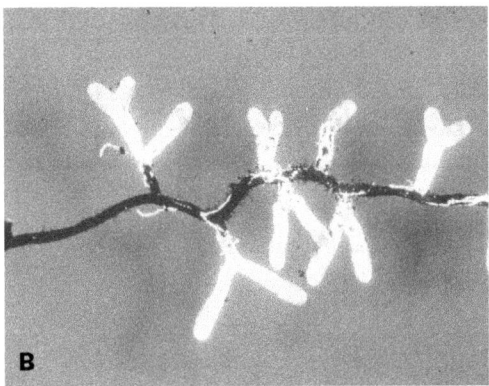

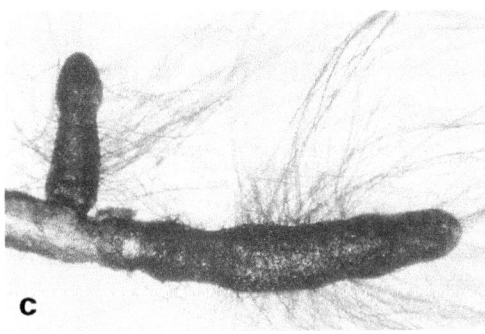

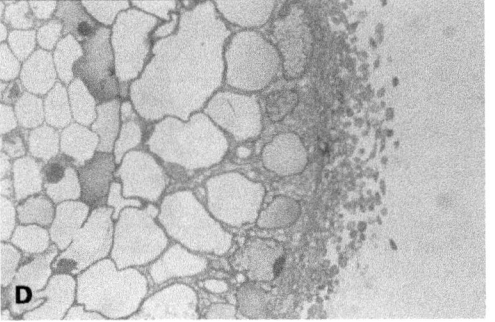

Figure III-5. Plant-fungal symbioses: ectomycorrhiza. A, An exceptionally highly branched pyramidal ectomycorrhiza formed on a balsam fir root. B, Dichotomic ectomycorrhizas formed on a pine root. C, Simple ectomycorrhizas formed by *Cenococcum geophilum*, an ascomycete. D, Anatomical details of an ectomycorrhiza showing the fungal sheath, the Hartig net, the cortical cells, and the endodermis. E, In boreal forest nurseries, container-grown seedlings often produce a profusion of ectomycorrhizal fungal fruit bodies in the fall. The photo shows *Laccaria bicolor* produced in association with balsam fir. Photo: A and D, J. André Fortin, Université Laval; B, Larry Peterson et al., CNRC; C, Linda Tackaberry, University of Guelph; E, Mohammed Lamhamedi, MRNF. From *Les mycorhizes. La nouvelle révolution verte*, by J. André Fortin, Christian Plenchette, and Yves Piché, Éditions MultiMondes, 2008. (Courtesy Jean-Marc Gagnon, Éditions Multi-Mondes, Québec, Québec, Canada)

Knudson and others. Growing orchids without fungus, under exceptional nutritional conditions, is comparable to the free culture of animal tissues in a corresponding nutrient medium.

When an embryo is still small and undifferentiated, the synthesis with fungus happens. The location where the fungus penetrates embryonic tissue ("transitional cells") is predictable.

The fungus usually is located completely inside the orchid, where it leads an intracellular life.

The formation of tubers, a characteristic organ widely found in orchids, was caused by the activity of symbiotic fungi.

In a tropical orchid *Didymoplexis,* in the absence of fungus, the roots are filamentous; the fungal infection leads to tuber formation (Magnus [1900]). MacDougal [1899] induced tubers in a completely different orchid, *Aplectrum,* through fungal infection.

Formation of a peculiar underground "nest," a "root" of *Neottia nidus-avis,* also appears to depend on the fungus's presence. Under the influence of symbiosis, lateral roots have exogenic origin instead of the normal endogenic one in this orchid.

In typical cases, like in *Neottia,* mycorrhizae are located in three to five cell layers (zones) from the root surface. Only in the fourth zone may mycorrhiza fully and indefinitely develop. In the third and fifth zones it forms, almost in every cell, amorphous thick globules, which eventually are digested by the cytoplasm of those cells ("digestive cells" of Magnus, "phagocytic cells" of other authors). In the cells of the fourth zone, hospitable to fungus, one can distinguish circular hyphae located at the periphery of a cell, and hyphae that serve for feeding. When the tuber dies, only the circular hyphae remain and preserve the fungus.

Inside an orchid, mycorrhizae find not only shelter but also food (water, carbohydrates). They process humic substances that are osmotically absorbed by the root. Orchids receive protein reserves by digestion of fungal "globules."

The systematic position of orchid fungi (or fungus) still awaits clarification. Experiments on isolation and culture with special media, and experiments on the synthesis of cultured fungi with sterile orchids, have been conducted with reasonable success. Orchid fungi were placed into the artificial genus *Rhyzoctonia* (which also includes a number of parasites) by Bernard [1909]. The same mycorrhizae were assigned to a special genus *Orcheomyces* by Burgeff [1912].

A fungus–orchid symbiosis was described by Kusano (1911). One of the symbionts was a saprophytic, rootless orchid, *Gastrodia elata*. Its partner was [the honey mushroom] *Armillaria mellea*, a very common, facultatively parasitic fungus.

Symbiotic relationships affect certain developmental stages of the consortium. During its vegetative period, the orchid does not depend on symbiosis. However, formation of the "stem mycorrhizae" is required for the flowering stage, not achieved unless the fungal symbiosis has been established.*

The fungus is necessary for the orchid's existence. This fungus can be in charge of root, tuber, or flower formation, and it can even modify an endogenic organ into an exogenic one: it can give stem characteristics to a root!

One must agree with Famintsyn: "It could be that we never knew and never saw any orchids, since what we call by this name is a product of symbiosis between a flowering plant and a fungus."

7. Heathers and Their Roots

Common heather, and probably other Ericaceae, is in fact the sum of a flowering plant and a fungus. This remarkable fact was demonstrated by Rayner (1913, 1915, 1916, 1921, 1923).

This fungus *(Phyllophoma)* is related to the genus *Phoma*, which includes fungi [Ascomycota] parasitic on plants.

In the seeds of heather *(Calluna vulgaris)*, mycorrhizae are located only in the cover layer. The embryo is infected through tender, transparent hyphae that penetrate from seed cover to the tip of the embryonic root.

If artificially prevented from this infection, the root of the embryo does not develop. Rootless plants, of course, are completely nonviable under ordinary conditions. Their life can be supported by rearing them on nutritive substrate, but even then, signs of degeneration are exhibited: yellow coloration or absence of color altogether, and their life span is short. In a sterile embryo

---

*Endotrophic mycorrhiza was likely present already in the first terrestrial plants, such as Devonian Psilophytales (compare recent data of Knudson, etc). [Currently, the earliest known mycorrhizae are indeed Devonian, ca. 400 mya (Taylor and Krings 2005).—*Eds.*]

infected with pure fungus culture (Rayner), the shoot acquires the ability to develop normally.

The fungus is necessary for the embryo's development, especially the formation of its root and its photosynthetic pigment. The heather root is a *symbioorgan,* i.e., an organ created by symbiosis. Since this organ is created by the activity of the fungal partner, it could be called a *"mycetome,"* or fungus-organ. We have already seen another fungus-organ, e.g., a flower or *Gastrodia* [orchid]. More fungus-organs will be described below.

In embryos that come into contact with fungi, the fungal hyphae easily penetrate through cell walls. Hyphae branch in the internal portion of the root without any resistance. A normal root develops, as well as a significant number of thin lateral roots where mycorrhizae send their offshoots. Mycorrhizae especially accumulate in the surface cells. In the stem the mycorrhizae penetrate all tissues at random.

Both shoot and adult plants lack root hairs.

The surface of the entire plant is very thinly coated by a fungal web, a continuation of the internal mycorrhizae.

Inside the root, localization of mycorrhiza persists in the adult as well. In the adult stem, mycorrhizae are found both in the bark and in the core. The most developed hyphal complexes are associated with the vascular bundles.

Mycorrhizae are abundant in the photosynthetic tissue of leaves. The hyphae extend to the plant's surface at random, but not through the stomata. In green cells, mycorrhizae come into intimate proximity with chloroplasts.

In flowers, fungus is present in all parts except the embryo sac and pollen.

The benefits for a fungus of life inside a plant are obvious. The benefits of the fungus for the flowering partner are less clear. The fungus may deliver atmospheric nitrogen to its partner in underground parts and in leaf tissue. [Fungi are eukaryotes; no eukaryote is known to fix atmospheric nitrogen; hence, this is unlikely.]

8. Toxic Glands of *Lolium temulentum*

Lolium temulentum, or darnel ryegrass, is a remarkable exception in general among grasses, which are normally nontoxic, and among its relatives such as rye, wheat, wheatgrass, etc.

The presence of mycorrhizae in darnel seeds, outside of the aleuronic layer, was discovered by Vogl (1898) and confirmed by other researchers.[12]

The mycorrhizae penetrate from the seed to the developing plant, where they reach newly formed ovaries and fruits. The fungus is inherited, and the toxicity of darnel seeds is related to the activity of these mycorrhizae. The fungus is necessary for the germination of darnel seeds (Freeman [1904]).

The darnel fungus belongs to Ustilaginales and represents a result of further degeneration of this fungal type (Brefeld [1872–1912]). Some ustilagos (*Ustilago tritici*, etc.) are transmitted through the seed, although they still retain reproduction through chlamydospores. These are absent in the fungal partner of darnel. A completely different opinion was presented (see p. 49) by Peklo [1913].

C. Third Series of Examples [Animals]

Examples 1–6 and 9–11 describe the best-studied symbiotic phenomena, those of insects; other examples refer to other animal groups.

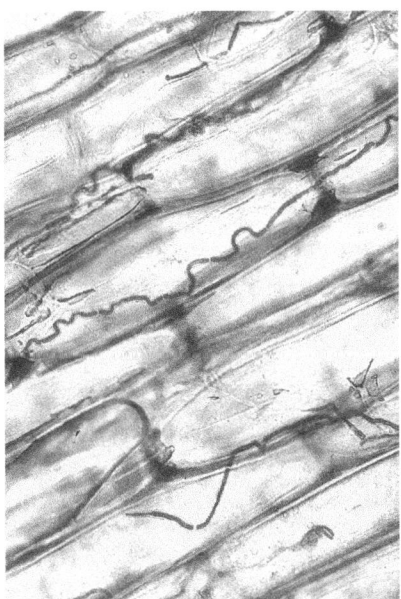

Figure III-6. Plant-fungal symbioses: endophyte mycelia. *Neotyphodium coenophialum* mycelia, which inhabit the intercellular spaces of tall fescue *(Festuca)* leaf sheath tissue. (Photo: Nick Hill [courtesy ARS USDA]).

1. Algal Pubescence in Dragonfly Larvae

Dragonfly larvae of the genus *Aeschna* can be covered by greenish pubescence [fuzz]. Although present all over their body, it is especially well-developed adjacent to the anus. After molting, it appears again on the new cuticle. Pubescent individuals appear more healthy and viable than those with poorly developed pubescence or lacking it. Pubescent larvae are larger than those with poorly developed pubescence.

This pubescence, formed by the green alga *Oedogonium undulatum*, is a symbiosis with a mutual benefit, suggests Kammerer (1907, 1908).[13]

The alga has the following benefits: (1) Better conditions for photosynthesis due to movement with the larva and production by it of carbon dioxide; (2) Availability of attachment sites on the larva's glabrous surface; (3) Protection from organisms that commonly consume algae.

The dragonfly larva benefits in that: (1) it lives in an atmosphere rich in oxygen; (2) it is protected by pubescence from parasites; (3) it is camouflaged by algae.

When the first "infection" of the larva occurs is not known.

2. Glandular Epithelium in Ant Midguts

In all species of the well-studied genus *Camponotus*, the midgut epithelium consists of two kinds of cells.[14] The midgut lumen is lined by large, swollen cells; they are followed by a layer of spherical cells with specific structure.

Bundles of long, thin "tubules" in those cells are found around an ovoid nucleus; these tubules can either be parallel to each other or form closed loops, like the Golgi apparatus. Sometimes, they fill almost the entire cell.

These "rods" or "tubules" were thought to be mitochondria by Strindberg [1913]. Blochmann (1884) and most recent authors, especially Buchner (1921), assert that these are peculiar symbiotic cytodes [bacteria]. The epithelium is a "symbiotissue," tissue created by the activities of symbionts.

The hereditary transmission of symbiotic cytodes, and therefore also the hereditary nature of symbiotissue, is provided by a very early delivery of symbionts to the egg. The symbionts must leave the gut epithelium and move to the egg chambers autonomously.

As the embryo develops, symbionts and the symbiotissue formed take a given position among other developing tissues.

That the insect and bacterium are mutualistic in their feeding physiology is confirmed by histological study of the ant midgut, as well as by the "effort," using Buchner's expression, that an embryo applies to deliver bacteria to their place.

These symbionts may provide the glandular function of the epithelium. The details of this process are known as little as that of any cell organelle, including the Golgi apparatus.

3. The Racemose Organ and Vaginal Glands of Beetles

The front portion of the midgut, on both sides, has blind saclike extensions in the beetles of the subfamily Anobiinae.[15] These are divided by septa into chambers and shaped like a grape bunch.

Undoubtedly, this structure is a digestive gland, where enzymes facilitate cellulose digestion. These beetles feed on plant material refractory to digestion.

The walls of this gland are built of two kinds of alternating cells: one ordinary kind similar to ciliated epithelium, and another of larger, inflated cells that lack cilia, between which the ciliated cells are located. These inflated cells, which in essence compose the entire organ, are replete with symbionts. Cultured separately by Escherich [1900], they are probably yeast.

The racemose organ, created by symbiosis, is a symbioorgan. Its enzymatic activity is due to the microorganisms within it.

Symbionts, first noticed here by Karawajew (1899), were claimed to be pathogenic [protists] flagellates. The situation was clarified by Escherich [1900], but scientists neglected his opinion. The question was ultimately solved by Buchner (1920, 1921), who confirmed the symbiosis.

The racemose organ is physiologically connected to a couple of thin tubular glands, which open into the vagina. These tubules, in their turn, are symbioorgans packed with symbiotic yeast. The hereditary character of symbiosis is provided in an absolutely exceptional manner for insects (see below). The vaginal glands supply symbionts to the surface of eggs laid by the beetle. Newly hatched larvae chew through the egg shell and swallow the symbionts, which are required for racemose-organ development. After symbionts enter the digestive tract, they interact with its walls and lead to the formation of the racemose organ.

4. False Yolk (Pseudovitellus) in Aphids

The term "pseudovitellus"[16] was first used by Huxley (1858) for a specific organ in some insects, i.e., aphids (Aphidae), that has a certain morphology and content, filled with yolk-like globules. In aphids, this organ consists of two halves that extend along the body and lodge in the abdomen, between respiratory muscles of segments 1 to 6. There both halves join above the hindgut. The surface is lined by flat cells; inside, cells are larger. The organ is supplied by tracheae of various complexity levels. Besides nuclei, internal cells have a high number of bodies similar to mitochondria.

Huxley and some others thought that this organ contains true yolk as a reserve. By contrast, Balbiani [1869–1872] concluded that this organ is homologous to testes; its existence in females indicated that the ancestors of aphids were sexless organisms.

The fact that males, in addition to "pseudovitellus," have true testes, Balbiani explained as follows. He suggested that these testes originated from female reproductive organs, i.e., are not real testes. Balbiani derived this conclusion from embryological, in particular topological, investigations. He also stated that pseudovitellus was a true organ, equivalent to the heart, liver, etc., and homologous to testes.

The pseudovitellus was considered by Witlaczil [1882] to be a kidney that substituted for Malpighian tubules.

Nobody doubted that the pseudovitellus is a normal organ until 1910, when simultaneously two scientists, Pierantoni and Šulc unexpectedly proposed a new interpretation. They showed that pseudovitellus is a symbiotic organ, or "mycetome," i.e., "fungus-organ," as they call it (classifying bacteria, according to old views, as fungi). "Mitochondria" mentioned above are, in fact, intracellular symbiotic cytodes.

The aphid symbiosis is inherited, and therefore the symbioorgan is also inherited. The hereditary transmission of symbionts is complicated by the alternation of generations in aphids. Delivery of symbionts to winter eggs happens easily due to the proximity of the symbioorgan to the reproductive organs. Symbionts enter through the rear end of the egg. A greenish plaque, the primordial pseudovitellus, soon forms. Invagination begins from the same side, and the primordium is embedded inside the egg. Later, they take their characteristic position relative to digestive organs. In parthenogenetic eggs the "infection" differs.

Based on some special considerations, one should think that pseudovitellus, or an organ that preceded it in evolution, appeared prior to the typical aphid developmental cycle.

The function of pseudovitellus remained obscure until the systematic position and specific way of life of symbiotic bacteria was clarified. The symbionts that produce pseudovitellus are identical to the nitrogen-fixing bacterium, *Azotobacter*, or very closely related to it (Peklo 1912 [1916 in original text]).

This bacterium assimilates atmospheric nitrogen. Thus, pseudovitellus, according to Peklo (1912), is a "nitrogen organ," comparable to well-known legume nodules.[17]

Another type of "false yolk" is known in aphids [whiteflies] from the group Aleurodidae. Here, paired glands lie close to reproductive organs. Male reproductive organs are surrounded by "yolk" as by an epithelium, which has a duct; female racemose reproductive organs are layered by "yolk." Cells comprising this "yolk" have twice as many chromosomes as body cells of the same animal (Schrader [1921, 1923]). They also contain numerous, more or less ellipsoid structures, less transparent than the rest of the cytoplasm. They are similar to but larger than mitochondria.

The role of pseudovitellus in this case is also unknown, although this organ was described by 1868 (Signoret). This too is a case of a symbioorgan; "mitochondria" mentioned above are symbiotic bacteria (Šulc 1910, Buchner 1912).

The delivery of bacteria to the eggs is achieved by the entire cell separating from the "mycetome" (Buchner 1912). These cells penetrate between the cells of the follicular epithelium by ameboid movement, accumulate at one end of an egg, and then with invagination are carried inside, later taking a certain location during development.

Two alternative interpretations derive from these details. Either the symbiont, which forms the symbioorgan, here is not a cytode but a cellular organism as in sponge-algae; or symbiotic bacteria in a cell cause duplication of the chromosome number in its nucleus.

Either of these conclusions is theoretically quite important. The first conclusion could be confirmed by future experiment, and then cells of this symbioorgan cannot be considered anymore as cells of the "host" modified under the symbiont's influence. The symbioorgan in this case is a sui generis colony of cells, foreign to the rest of the organism.

5. Cicadas: Their Abdominal Organs

In Cicadidae, this organ[18] typically is an ovoid, more or less lobate body, which lies under the hypodermis, between the third and the beginning of the sixth abdominal segment, and is supplied by a large tracheal trunk. The organ is divided into segments or chambers of two kinds. External segments form a sort of covering and have a characteristic colored epithelium with particles of red pigment.

Internal segments lack epithelium. They appear to be accumulations of the cytoplasm with numerous and variable nuclei. Central, nonpigmented cells of external segments and entire internal segments include numerous sausage-shaped bodies. These bodies vary from peripheral to internal segments in both shape and staining reaction.

The organ, with its very complicated structure, was thought enigmatic. Some authors regarded it as a reproductive gland.

This is a symbioorgan, or mycetome. Two types of autonomous cytodes reside in it (Šulc 1910). One, whether bacteria or yeast, lives in the central cells of external segments; another inhabits the internal area of the organ.

This symbioorgan, as well as microorganisms that create it, is inherited. Regarding inheritance of the symbionts, according to the detailed studies by Buchner (1921), "both microorganisms, notwithstanding their difference from each other, penetrate the egg at the same time and in the same place."

In cicadas, these cytodes, by the time eggs develop, form a primordium that serves for hereditary transmission (Buchner [1921]). In *Cicada orni*, characteristic sausage-shaped bodies are replaced by ovoid or even almost spherical ones. The "infection" occurs relatively late, when separate cytodes leave their organ, and after a certain time as free-living microbes they penetrate through the follicular epithelium to the egg's posterior end. The development that follows resembles that of a real, "true" organ.

6. Lice: Their Hepatopancreas and Oviduct Ampullae

The "abdominal organ"[19] (which many authors, starting with Robert Hooke, identified as the liver), lies in the depression of the midgut. The area where the midgut is especially thick is a reservoir of absorbed blood. In the body louse, the organ is typically more or less ovoid and yellow and consists of two parts: first, a thick, sturdy syncytial covering that lacks nuclei. Second, a group of chambers (10 to 24 or more, depending on species) are located in a more or less horseshoe-like arrangement. In those chambers, no nuclei exist (or can be discerned), but instead there are a large number of filamentous inclusions. The covering of the organ forms layers between chambers and a central rodlike structure.

This organ was considered for a very long time to be a digestive gland. The filamentous inclusions are cytodes, similar to bacteria, and the entire organ is a symbioorgan, or mycetome, as demonstrated by Sikora and Buchner in 1919–1920.

Hereditary transmission of symbionts through eggs was already observed by Sikora. A special organ for this transmission, also known for a long time, is called an "ovoid ampullae" (Buchner 1920, 1921).

Each oviduct has an inflated portion, adjacent to the ovary. The wall of the inflations (ampullae) contrasts with the rest of the oviduct wall. They consist of three cell layers. One is comprised of cells in which the nuclei are located basally, in which a major portion of a cell is occupied by giant vacuoles. The vacuoles house large, sausage-like symbionts.

Further posteriad, the ampulla grades into a standard oviduct "just like *porto vaginalis* in a human uterus grades into the vagina."

Eggs are conveniently supplied by the symbionts as they pass the oviduct. During embryonic development, the primordium of the symbioorgan, in its specific location, passes through a number of stages. It resembles other organs.

7. Ticks: Their Digestive Glands

In a tick *Liponyssus saurarum,* which feeds on lizard blood, the midgut, as in all arachnids, is equipped with blind saclike diverticula. The midgut has only one epithelial layer. A muscular cover encloses not only this layer, but also a pair of ovaries and three special digestive glands.[20]

Two of those glands are located symmetrically in the walls of the right and left midgut diverticula; the third is located on the ventral side of the middle (posterior) diverticulum, in a special cavity between muscles. The paired glands are ellipsoid, with their diameter equal to that of the blind diverticulum. The nonpaired gland is kidney-shaped and larger in size. Each gland is a complex of large cells.

Each cell [of these glands] under normal conditions is full of bacteria (Reichenow 1922). He distinguished four kinds of cytodes that differ in shape, size, and structure. One that resembles contractile fibers was described above (chap. II).

These glands very closely resemble the digestive apparatus. They are inherited due to transmission of symbiotic cytodes from generation to generation.

A fertilized egg, before it passes to the uterus, is located in the body cavity right between two lateral mycetomes and under the median one. Therefore, transmission of cytodes is fully secured. By the time an egg reaches the uterus, the bacteria are already scattered deeply inside its yolk. The glands are formed at the same time as gut epithelium.

Organs similar in structure, role, and symbiotic nature are known in various other ticks, where they may vary in number and position. Their structure may be more complex and involve the tracheal system.

8. Esophageal Glands of Leeches

In a leech *Placobdella catenigera,* and probably in other species, including the medicinal leech, bottle-shaped organs open into the esophagus behind two pairs of salivary glands.[21] They differ from salivary glands in their location and size; A. Kowalewsky named them "Oesophagusdrüsen" [esophageal glands].

"In very young leeches that still live on yolk, the primordia of those glands appear as two cylindrical extensions of the esophagus. Their cells are still of the same size and shape as wall cells of the esophagus. In adult leeches, shortly before the first blood meal, these cells acquire a glandular character. Some epithelial cells increase in size, others remain between them as septae. Inside the enlarged cells high numbers of filamentous bodies are seen. The entire cytoplasm seems to consist of them (fibrillar, or filamentous cytoplasm structure!). The filaments are relatively long, stained weakly with hematoxylin, and contain no chromatin. They are so thin that they can be distinguished only at the highest magnifications. They often extend to the lumen of the gland. A leech examined soon after its blood meal shows these filaments in great numbers in the anterior of the digestive tract" [Kowalewsky 1901].

Contents of the glands are emptied into the stomach after feeding, and the filaments can be seen in liquefied food, especially in the posterior portion of the gut. They appear either separately or in globules. Ingested blood cells next to them are digested to a homogeneous mass. The role of these filaments seems to be to digest food cells.

"Without a doubt" these filaments are cell products (Zellprodukte) (Reichenow 1910). On the contrary, Siegel [1903] assumed they are hemogregarine parasites. He found them even in the youngest leeches and was convinced of a "hereditary character of the infection."

Repeated and new investigations of Reichenow forced him to reject his former opinion. Siegel was closer to the truth.

The filaments are independent cytodes, and the "Oesophagusdrüsen" are typical symbioorgans (Reichenow 1922). Their function is performed by microorganisms that create and inhabit them.

9. Lepidoptera: Accessory Glands of Their Reproductive Organs

The accessory glands were interpreted to be cell organelles, crystalloids, or "false nuclei." They resemble pebrine [*Nosema* fungus] as well as yolk plates of some fish and amphibians.

They are symbiotic cytodes: a bacillus similar to *Bacillus mycoides* and yeast. Other partners may also be present.

Therefore, the fat body in Lepidoptera is a mycetome, or symbioorgan. The reproductive organs develop from penetration of symbiotic cytodes and excreta from this accessory gland. The symbioorgan is a part of the Lepidoptera reproductive apparatus.

Prior to degeneration of the fat body, the symbionts are at rest, enclosed in a covering. The effect of migration on the fat body is the effect on symbionts; it causes disappearance of their covering and makes them more motile.

The hereditary transmission of symbionts from generation to generation is secured by their penetration into reproductive organs. Specific infecting glands, so-called lubricating glands, were found in the beet webworm [*Loxostege sticticalis*, Pyralidae] (Pospelov 1922). They form a pair of rather long and thin tubules that open into the posterior end of the female oviducts. The tubules are replete with cytodes that resemble yeast. These symbioorgans ensure hereditary transmission of symbionts.

During embryonic development, the primordium of the symbioorgan, as in other cases, takes up a given location among growing tissues to resemble a "true" developing organ. Nobody suspected the true nature of the fat body for a long time.

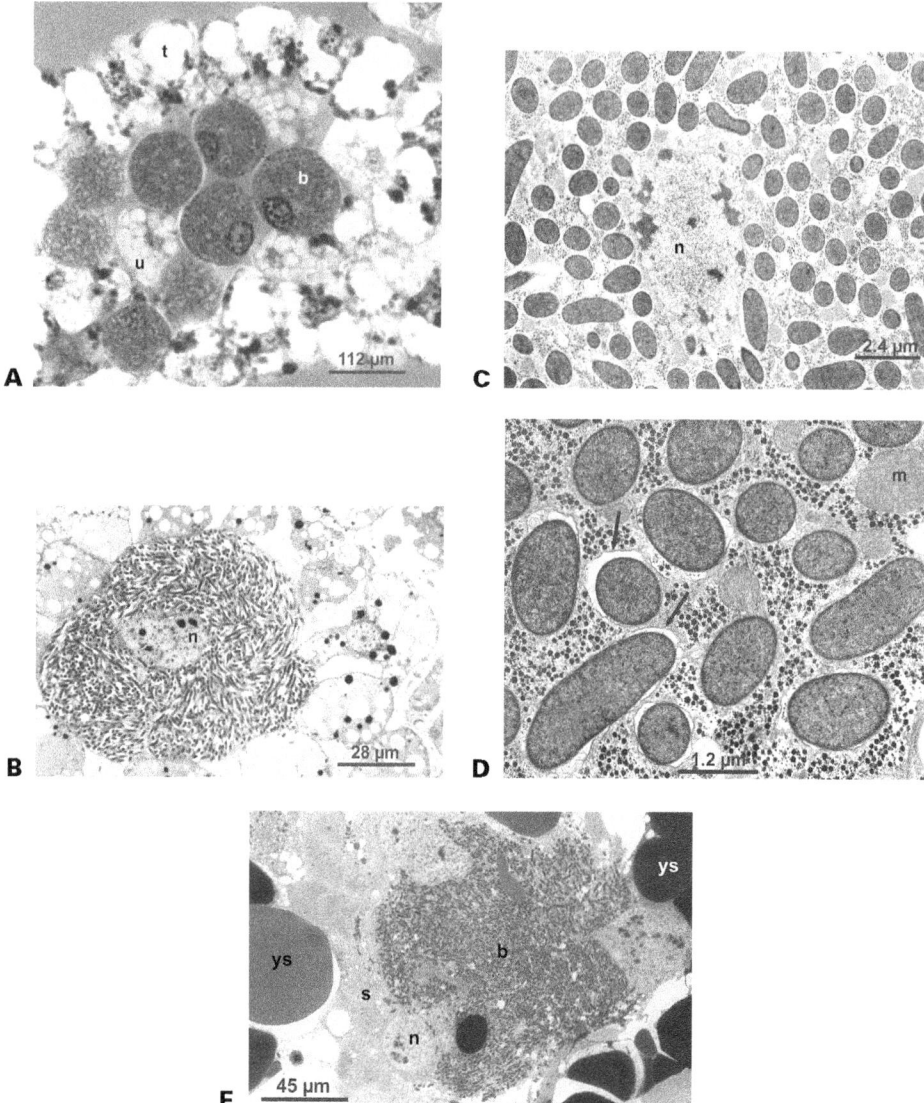

Figure III-7. An insect-bacterial symbiosis: symbiotic bacteria *(Blattabacterium)*, which form "mycetomes" in cockroaches. A, fat body of *Blattella germanica* (b = bacteriocyte; u = urate cell; t = trophocyte). B, bacteriocyte of *B. germanica:* the cytoplasm filled by symbiotic bacteria (n = bacteriocyte nucleus). C, bacteriocyte; n = nucleus. D, details of the bacteriocyte cytoplasm. Symbiotic bacteria are encircled by a vacuolar membrane (arrows) (m = mitochondrion). E, mycetome in the middle region of the yolk sac of *Periplaneta americana:* a cluster of bacteria (b) surrounded by a syncytial envelope (s) (n = syncytial nucleus; ys = yolk spheres). (TEM micrographs, courtesy Luciano Sacchi, Università degli Studi, Pavia, Italy)

The activity of reproductive organs depends on the presence and activity of symbiotic cytodes (Pospelov 1922). The symbiotic cytodes are regulated by the behavior of the insect, i.e., increased movement, change of feeding regime, etc.[22]

Symbiotic cytodes thus not only influence sexual maturation of Lepidoptera, locusts, and other animals but also may cause their migrations and underlie their specific "instincts." They thus stimulate their ravaging invasions of our fields and meadows.

The fat body of cockroaches is also well-studied.[23] Only the differences from the fat bodies described above will be mentioned.

The symbionts [of cockroaches] belong to the group that possesses a convoluted flagellum, named *Bacillus cuenoti*. They were considered to be a product of metabolism for a long time (Cuenot, Prenant, Henneguy) or to be mitochondria (Schneider).

The symbionts of a common cockroach were successfully isolated and reared in pure culture (Mercier 1906).

Not all fat body cells include symbionts. The position of "mycetocytes," cells that house cytodes, is not identical in different species. "Infecting" glands are not known.

A constant and repeated finding of bacteria in cockroaches inside a special organ, on the one hand, and the well-being of "infected" cockroaches, on the other, leaves no doubt that this is the case of mutualistic symbiosis. However, the role of symbionts and symbioorgan is not yet known. The fat body may participate in food processing, Buchner [1921] suggests. After the works of Pospelov [1922], however, it seems more probable that the role of the fat body here is similar to that in Lepidoptera.

10. Bedbugs: Their Paired Glands

The paired incretory glands[24] exist in both genders: in males, the gland is an ovoid, distinct body that lies in a depression of the testis. The position of the glands is less definite in females; to see them one needs to remove the fat-body lobes.

The size of a single gland is 0.3 to 0.5 mm. A group of large cells is enclosed in a thin covering that branches inside the organ. A system of tracheae

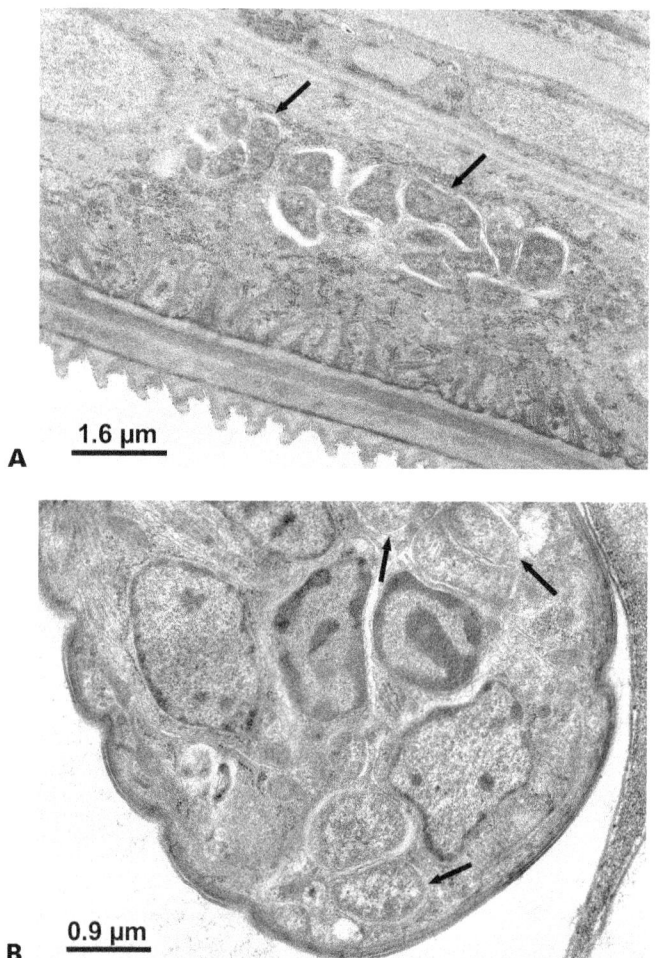

Figure III-8. An animal-bacterial symbiosis: symbiotic bacteria *(Wolbachia)* in filarial nematodes (*Brugia pahangi,* human endosymbionts, cause of "river blindness"!). A, lateral hypodermal cord of the lymphatic filaria with clusters of *Wolbachia* (arrows). B, lymphatic filaria with a few *Wolbachia* (arrows). (TEM micrographs, courtesy Luciano Sacchi, Università degli Studi, Pavia, Italy)

penetrates inside. Besides their giant size, the large cells are remarkable in that they contain three to five nuclei each and numerous inclusions similar to mitochondria.

The role of these paired organs is unknown. The organ may be incretory, related to food digestion by the bedbug, wrote Buchner [1923].

These glands also are symbioorgans, or "mycetomes" (Buchner 1923). Bodies similar to mitochondria also proved to be bacteria, quite variable in shape. Two different rodlike cytodes might be in them.

The presence of bacteria also in other bedbug organs, e.g., in sperm receptacles, was noticed by Buchner [1923]. Also they were present in "Ribaga's organ" [paragenital sinus, ectospermalege] (a special unpaired saclike organ, which lies in the left side of the abdomen and facilitates resorption of spermatozoa). Possibly, in this case scientists in the near future will also encounter new symbiotic phenomena closely related to reproductive functions.

11. Beetles' Light Organs

Paired organs on the penultimate abdominal segment are found in the larval stage of fireflies (Lampyridae), e.g., in *"noctiluca"* [*Lampyris noctiluca*].[25] They are retained through all molts. Larval organs remain only in males. Females add two small organs on the fourth abdominal segment, and two large unpaired luminescent plaques on the fifth and sixth segments.

Two cell types are distinguished in this organ. One forms an internal, nontransparent zone and is recognized by its large content of urates. This zone serves as a reflector. Another, comprised of luminescent cells, forms a transparent zone that faces the chitinous cover of the body. Its cells are filled by "tender granules" that Dubois [1914] considered to be plastids.

The flow of oxygen, of great importance for luminescence, is maintained by a well-developed system of tracheae in this organ. The tracheae enter the luminescent layer, where they branch into thin capillaries. Influx of oxygen is also maintained by blood circulation. If one presses a beetle's abdomen to increase hemolymph flow to luminescent organs, these organs flicker or increase in light intensity. If one blocks the hemolymph flow, the opposite occurs: the light dims. A direct connection between the "heart" contraction and light impulses in fireflies has been long known. The supply of oxygen through blood is a more primitive feature, according to Dubois [1914].

The described luminescent organs are, again, "mycetomes," as Pierantoni (1914) demonstrated.

Luminescent bacteria are the symbionts that induce formation of these organs. They were mistaken for plastids and other cell organelles by Dubois and others. Therefore, fireflies glow with the light of their symbiotic cytodes.

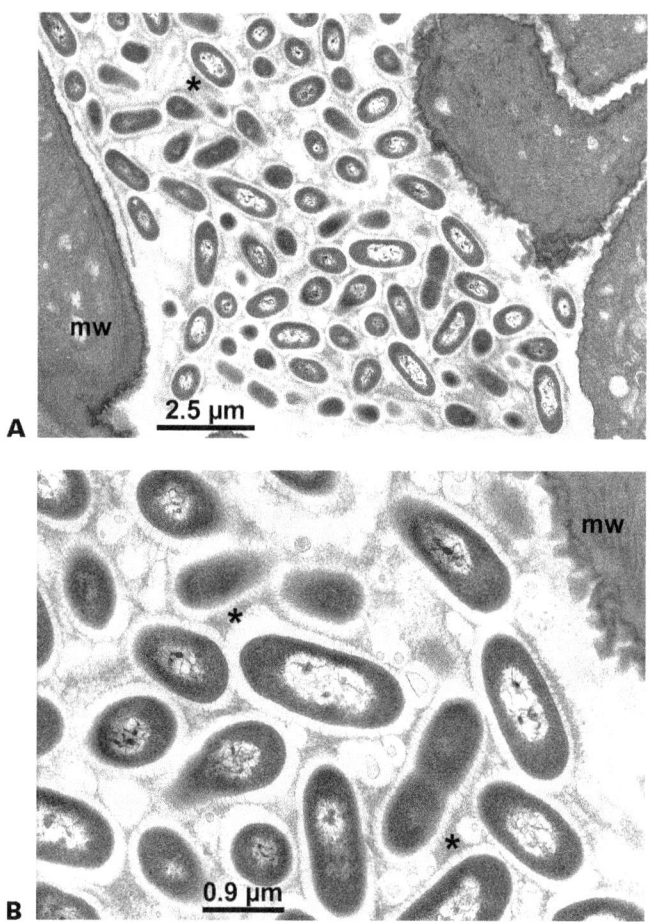

Figure III-9. An insect-bacterial symbiosis: Symbiotic bacteria *(Asaia)* in the midgut of the malarian mosquito *(Anopheles stephensi)*. A, Midgut lumen of an adult female of the mosquito vector *Anopheles*, full of *Asaia* (asterisk = extracellular slime matrix; mw = midgut wall). B, details of *Asaia* symbionts; note the presence of an extracellular slime matrix (asterisks), an electrondense cytoplasm, and a filamentous nucleoid region (mw = midgut wall). (TEM micrographs, courtesy Luciano Sacchi, Università degli Studi, Pavia, Italy)

The organ is built by two types of bacteria that were isolated in pure culture (Pierantoni [1914]). In artificial conditions they appear to lose luminescence.

Luminescent bacterial parasites of isopods also fail to glow in culture. But if they are transferred from culture to the isopod, luminescence is restored.

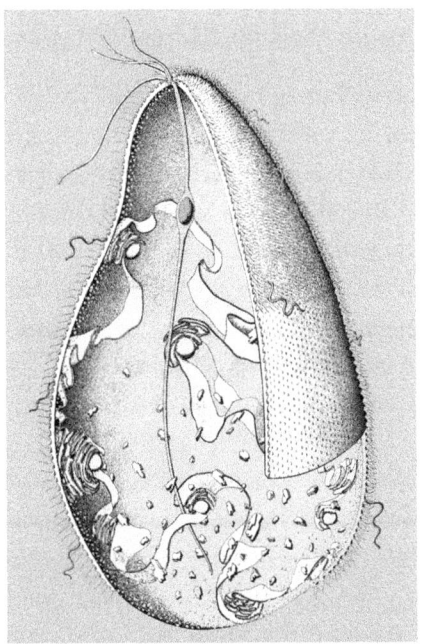

A

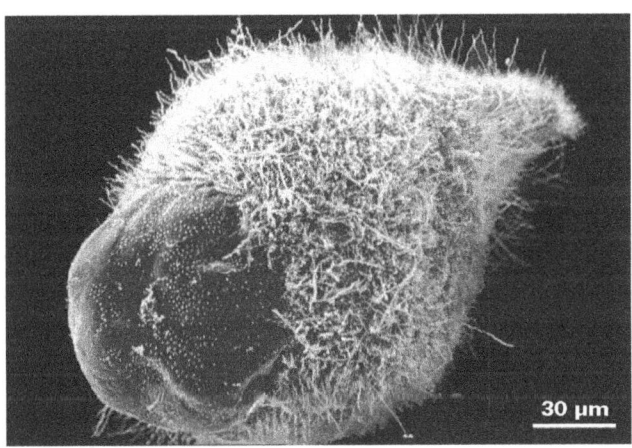

B

Figure III-10. An insect-protoctist-bacterial cellulolytic community inside the Australian termite *Mastotermes darwiniensis*. The hypertrophied trichomonad *Mixotricha paradoxa*. In this amitochondriate protoctist (Archaeprotista), thousands of treponema-like surface spirochetes are embedded in the outer layer of the cell (cortex) (Wier et al. 2010). By their coordinated swimming movements, these thin symbiotic eubacteria propel the organism in an anterior direction. *Mixotricha paradoxa* ingests wood particles through phagocytosis, aided by different (borrelia-like) spirochetes. A, analytical drawing by Christie Lyons based on videos, transmission, and thin-section electron micrographs such as the whole-cell *M. paradoxa*. B, SEM micrograph (bar = 30 μm).

12. *Cyclostoma elegans,* "Storage Kidney" (Mollusca)

In connective tissue of this terrestrial gastropod, in a dorsal location between kidney and stomach, lies a multicellular body that functions as a storage kidney.[26] It is a true glandular organ, which has main and lateral saclike portions, with offshoots of connective tissue between them. Numerous lacunae and blood vessels penetrate this volume.

Cell structure in the mature storage kidney is specific. When the organ is young, small, dense globules in some of the vacuoles are noticed. They have a variable number of "nuclei" and several cover layers. As secretion increases, these globules fuse and form common cover layers. Around these globules, cytoplasm contains a very large number of "rodlike bodies" from 3 to 5 µm that quickly reproduce. They were considered to be either parasitic bacteria or phosphate mineral deposits. They tend to join into chains, filaments, etc. Growth of "globules," reproduction of rods, and increase of cell size occur at the same time. The globules contain uric acid and xanthine, so here we deal with a true kidney (Mercier 1911–1913).

When the accumulation of excreta reaches its limit, amoeboid cells enter. They are found in large numbers in adjacent lacunae and vessels. The amoebic cells are phagocytes that participate in digestion of the "globule."

Columns and rods in these kidney cells are autonomous cytodes: bacteria. The kidney itself is a mycetome, or symbioorgan, as demonstrated by Mercier.

The symbionts may be transmitted through generations with egg "infection."

This symbiosis may be based on the ability of bacteria to utilize excreta collected in this location for protein synthesis. With bacterial help, the mollusc removes its metabolic products and also receives extra food by consumption of the "bacteria" in its cells. The parallel to the association of worms and algae was described above (see pp. 21–24).

13. Tunicate: Its Bojanus Organ [Renal Sac]

The Bojanus organ in the Molgulid tunicates is located in the right half of the body, near the heart. It lies in a fold of digestive tract, near the reproductive gland. It lacks any outlet and is shaped like a cylinder with rounded ends, slightly flexed like a bean. Its walls consist of a muscular layer lined inside by a single-celled epithelium.

This organ contains many strange inclusions, as has been known for a long time. They were called either confervoid filaments (i.e., similar to *Conferva* [green] algae[27]) or parasitic gregarines.

This kidney is a symbiotic organ,[28] a mycetome with diverse population, according to Giard [1888] and Hellmann [1913].* It is a shelter for a highly organized chytrid fungus, according to Giard [1888], who discovered this fungus and named it *Nephromyces.*

This fungus entangles concretions in the kidney with its mycelium. Some of its branches are thick and form organs of zoospore production. Zoospores, spherical with a single long flagellum [undulipodium], were noticed. Sexual processes were observed that lead to the formation of zoospores with rough coverings.

Besides the sporangia of this enigmatic microbe, Hellmann [1913] found a significant number of previously unknown spirochetes.

Whether the kidney includes several heterogeneous symbionts or just one but with great polymorphism is unknown. If the latter, a single yeast "fungus" at different developmental stages might resemble a fungus, a spirochete, a bacterium.

The tunicate benefits from the presence of this fungus, because it removes excreta accumulated in a completely closed organ (Giard [1888]). The tunicate later partially consumes the symbionts, apparently, because otherwise they would overfill the organ.

14. Luminescence in Tunicates: *Pyrosoma*

The tunicates of the genus *Pyrosoma* emit a magnificent glow under various stimuli.[29] The glow forms a wave starting from the point of stimulation. Alcohol and ether increase the intensity of the light; carbon dioxide poisoning leads to loss of response. "It is possible that the luminescence depends on separate members of the colony" (*Pyrosoma* forms tubular colonies).

The entire colony does not glow; rather, certain points are luminescent. In a colony 8 cm long, up to 6,500 luminous points can be counted. Each individual tunicate in the colony has two luminescent organs. These organs, two

---

*For information on this work, which was overlooked in Buchner's review, I am greatly obliged to Prof. K. K. Saint-Hilaire.

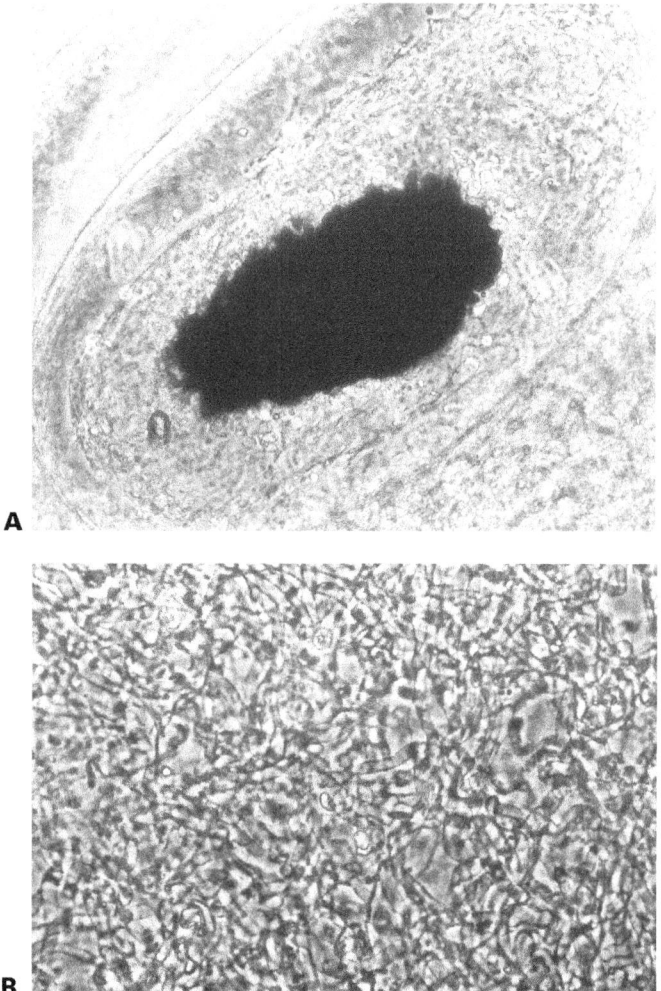

Figure III-11. An animal-protoctist-bacterial symbiosis: *Nephromyces* in the ascidian tunicates *Molgula*. A, *Nephromyces* surrounding concretion in the renal sac of *Molgula manhattensis*. B, Dense concentration of vacuolated filaments of *Nephromyces* from the renal sac of *M. occidentalis* (Phase contrast; photo: Mary Beth Saffo, Marine Biological Laboratory, Woods Hole).

mesodermal groups of cells, are located laterally from the "gullet," in the peripharyngeal blood sinus.

Each luminous organ is a one- or two-layer complex of a variable but large number of cells. Each cell has a distinct, constant nucleus and harbors many "grains" that resemble oil drops. Luminescence, some suggest, is caused by slow oxidation of these drops. The drops are more or less rectangular. Joining each other, they form chains and globules. Inside them, even smaller granules that stain like chromatin are seen. Drops or "grains" themselves stain similar to lignin.

The luminescent organ tends to a round or ellipsoid shape, or it may be shaped as a thread or a band. Besides the two "mandatory" organs, another pair may be present near the cloacal opening.

Early authors knew about these organs. Some (e.g., Savigny) considered them to be ovaries, while others, including Huxley, claimed they are kidneys.

Regarding the nature of "granules" or "drops" with their "little granules" inside, opinions differ. The drops were thought to be cytoplasmic structures, and their granules mitochondria, by Julin [1912]. Each globule, an association of granules, Kowalewsky considered a true nucleus. That an entire cell contains such globules was taken to indicate a conglomeration of cells poor in cytoplasm.

The enigmatic inclusions of the luminescent organ are autonomous organisms (Buchner 1914). The globules and chains are their colonies. The organs themselves are symbioorgans, or mycetomes. Luminescence is produced not by the *Pyrosoma* animal, but by the symbionts. Their presence and activity created these luminescent organs. This symbiosis is hereditary: symbiotic cytodes are probably carried to egg chambers by blood current. Young luminescent cells are very similar to blood cells.

During embryonic development, the cells that include the cytodes disperse among blastomeres. The short-lived larva, a "cyathozoid," lacks typical concentrated luminescent organs, but it also has no muscles, gut, gills, and other organs. How luminescent organs develop in adult "ascidiozoids," which asexually appear on a cyathozoid, is yet unknown.

15. Cephalopod Nidamental Glands and Light Organs

Luminescent organs of cephalopods[30] are known in the Dibranchia group, and in particular in Decapoda [Decapodiformes: squids and cuttlefish]. These organs are connected to accessory nidamental glands.

In myopsid decapods, true nidamental glands are present only in females. Paired, pear-shaped sacs, located symmetrically, they accompany the ink sac. The folded connective tissue and ciliated epithelium, which lines it, create the leaflike structure of the organ. The glands secrete substances that, together with secretions by other glands, are used to form the egg cover.

These organs are joined by accessory nidamental glands that lie close together and are fused. Until recently these organs also have been considered to be organs used for the development of the egg cover.

In *Loligo forbesii* such glands are present in males but are quite underdeveloped (Wülker 1913). These organs do not contain any secretions, but instead the cells lining these glands are full of small, rodlike bacteria (Pierantoni 1918). That "accessory glands" are a type of symbioorgans was demonstrated by Pierantoni.

The role of these glands remains unclear. No relationship with luminescence was found. The fact that they are underdeveloped in males might mean that they are used only by females.

In species where accessory glands are only in females, the situation is different. *Sepia elegans* (the cuttlefish) has two round glands with radial folds that enlarge toward the periphery. Tubular cavities here spread deeper than in *Loligo*. They converge and diverge to form an internal labyrinth. The organ therefore becomes more compact and increases in size. Openings of tubes converge in its middle zone. Here, symbiotic bacteria are also present in great numbers. In species of the genera *Sepia* and *Sepiola*, hereditary transmission of symbionts was demonstrated. The same probably occurs in other species.

A most interesting feature of *Sepia elegans* is the presence of three types of tubes of different coloration. They are populated by different bacteria. Tubes can be either whitish, with short, slender rodlike bacteria; or yellow, with short rods; or red-orange, with very small spheres (cocci). Bacteria retained their characteristic features in pure culture obtained by Pierantoni. "A remarkable fact discovered then was that only symbionts from yellow sacs were able to glow in the dark" (Buchner [1921]).

The accessory nidamental glands, partially, are luminescent organs. Their activity assures that the ventral body surface in the female glows during its mating period, and to a lesser degree, at other times. The position of luminescent glands secures hereditary transmission of symbionts from the female parent.

Further specialization of the symbioorgan can be illustrated by *Rondeletiola minor* [family Sepiolidae]. At first glance, this species has one unpaired luminescent organ that lies on the ink sac. This organ, in fact, is paired, but its two parts are close together. Accessory nidamental glands join it on both sides.

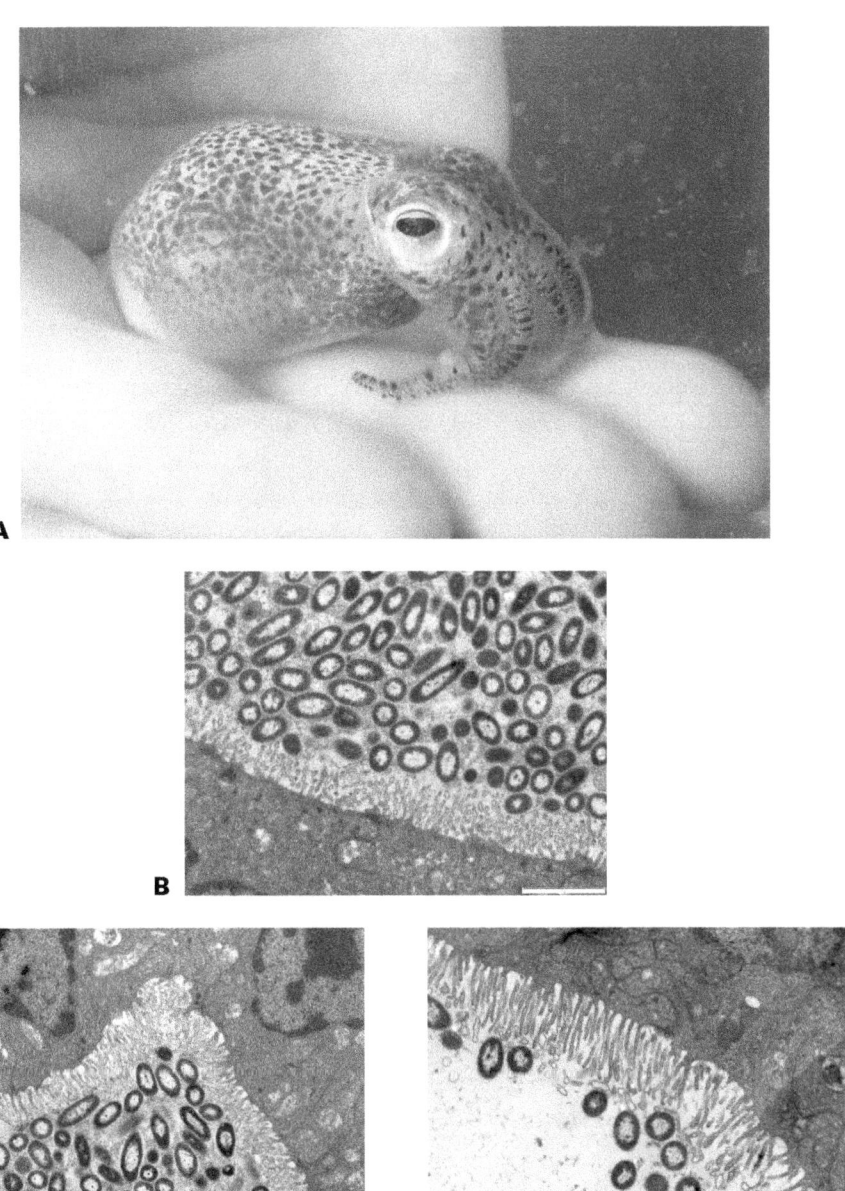

Figure III-12. An animal-bacterial symbiosis: *Vibrio fischeri* in the light organ of *Euprymna scolopes,* the Hawaiian bobtail squid (Mollusca: Cephalopoda). A, squid. B-D, *Vibrio fischeri,* TEM micrograph (bar = 2 μm), (photo by Ben August, courtesy Andrew Wier, University of Wisconsin, Madison).

The nidamental glands, as well as the luminescent organ itself, form a collection of sacs or tubes. Only sacs of the glands are white, while those of the organ are yellow. Orange sacs form only a small fringe around the luminescent organ. The latter lies inside a depression of the ink sac.

The entire structure is equivalent to the nidamental organ of *Sepia,* with the only difference that the luminescent portion is more independent. The luminescent part is also present in males, where it is even more developed.

Luminescence here is also caused by bacteria, in this case of two sorts: cocci and rods (Pierantoni [1918]).

With modification of the luminescent portion into an independent organ, specific adaptations appear that increase the activity of symbionts. Bottle-shaped tubes, which increase in diameter downward, are surrounded by connective tissue remarkably rich in capillaries that feed the symbionts. Connective tissue on the top is transparent; its differentiation into a lens can be seen. A reflecting mass is formed on the opposite side, the ink sac located below it as a nontransparent screen.

In *Sepiola elegans,* as studied by Pierantoni, two symmetrically positioned organs are not as close to the accessory glands, but still they are adjacent to the ink sac. The accessory gland is an unpaired, round organ; its large central portion is built by orange-red sacs full of bacteria. White sacs lie only on the circumference. The orange sacs contain short rods, whereas the white ones contain cocci.

Both sorts of bacteria retain their features in pure culture. The luminescent organs, just as in *Rondeletiola,* are inhabited by swarms of nearly spherical cocci and long rods.

A transparent lens is located above dense populations of symbiotic bacteria. Due to this circumstance, exit openings of the sacs are shifted to one side. A "tapetal" lining layer is well developed on the bottom side of the organ. The ink sac in this case also serves as a nontransparent screen. It is inserted between lens and luminescent mass and thus assumes a cuplike shape. Only a small opening remains from which light is emitted. Thus, and because of the contractibility of the reflector layer, the light is regulated voluntarily. The hereditary transmission of symbionts is achieved through eggs. The structure of this organ resembles that of an eye.

The most specialized luminescent organs apparently exist in deep-water oligopsids, where they are found in large numbers and tend to be present all over the body. Due to various modifications, the emitted light can change color.

Instead of folds with "sacs," these organs form isolated cellular masses. Symbiotic cytodes are located between, rather than within, cells. The cellular masses are supplied by a well-developed network of capillaries, so that the luminescent organ has a spongy appearance. The structure of the organ is further complicated by branching nerves. Nothing is known about the inheritance mechanism. In the early embryo, symbionts are scattered; later, they give rise to, or trigger, organ formation at those points most suitable for development of the bacteria.

Two explanations for the origin of these structural phenomena were proffered by Buchner [1921]. The symbionts that had no connection to luminescence were at first present only in females, located in the accessory nidamental glands, and played some role in reproduction. The presence of such glands in males should be regarded as a "consequence of heredity." Luminescent symbionts joined this association only later.

It is also possible that the accessory glands originally served for the transmission of symbionts, comparable to "lubricating" glands of Lepidoptera, which had *another*, permanent or main location. In this case, accessory nidamental glands in the male of *Loligo forbesii* are a useless, and thus underdeveloped, organ of transmission.

16. Blood Platelets (Hematoblasts) in Mammals[31]

Hematoblasts are found in enormous numbers in blood plasma. These are transparent, more or less rodlike structures, ca. 3 μm in diameter. Their number is proportional to the intensity of metabolism. They exhibit the same reactions as mitochondria. They have independent movement, typical for cytodes. Starving causes specific changes in hematoblasts.

Hematoblasts generally are considered cell components of blood, along with red and white blood cells, erythrocytes, and leucocytes, and equivalent to thrombocytes of amphibians, fish, etc.

Hayem [1877] considered them to be cells destined to develop into erythrocytes. This opinion was demonstrated to be incorrect.

Aynaud [1909], who specifically studied hematoblasts, concluded that their role is absolutely unknown. He states that they are remarkably similar in structure, staining, activity, etc., to bacteria; one could suggest that these are not blood elements, but symbiotic bacteria.

This view was energetically supported by Portier (1918). It agrees well with the experimental results of Courmont and André, who cultured blood platelets.

D. Other Tissues and Organs

We have listed a large number of examples, present in a large variety of organisms, of tissues and organs that differ greatly in their structure, location, and function.

These tissues and organs result from symbiogenesis; they are produced by the activity of heterogeneous, autonomous symbionts.

The examples here do not exhaust all symbiogenetic origins of tissues and organs known to science, of course. Only the most typical examples and, most importantly, only those which are not controversial are listed. Many *similar* examples are known.

To what extent are these tissues and organs, which proved to be symbio-tissues and symbioorgans, typical tissues and organs?

All symbiotic tissues and organs mentioned here were considered quite typical, "regular" tissues and organs. Until very recently this was the opinion of the best specialists.

Nobody suspected how widespread symbioorgans are, and the very term "mycetome" did not exist before 1910. The true nature of luminescent organs was not suspected until 1914. Just a few months ago the presence of symbio-organs was demonstrated for leeches and arachnids. One can say that just yesterday we learned about a remarkable connection between the "infertility" phenomenon in Lepidoptera and symbiosis.

Almost every new specialized journal publication brings information about the symbiotic nature of various tissues and organs in diverse organisms. The nearest future may confirm the hypothesis of Pierantoni [1910] that dyes of cochineal [*Coccus cacti*] and the purple dye of mollusks are produced by symbiotic bacteria, or the hypothesis of Buchner [1921] that poison organs of animals are symbioorgans, and that "animal" toxins are in fact toxins of sym-biotic cytode toxins. The "trabeculae of Sanio," a characteristic ornamentation of the vascular tissue in numerous plants, may be a product of fungal symbi-onts inside timber (the hypothesis of Jeffrey [1921]). A great ruckus arose around experiments of Portier [1918] and his followers, who discovered myr-iad symbiotic cytodes in the testes, pancreas, and other organs, not only in amphibians but also in birds and mammals.

Simple statistics convinces us that, if such discoveries develop at a rate characteristic for recent years, in the relatively near future the majority of tissues and organs in animals and plants will be transferred to the category of symbiogenetic phenomena.

"Recently," wrote Elie Metschnikoff [1909], "I visited experimental plantations near Paris, at Medon, maintained by Blaringhem with the goal of studying plant variation and origin of new varieties and species.

"As it was already observed and studied earlier, especially by De Vries, new characters started appearing right away in Blaringhem's cultivated plants. Especially clear changes appeared in stalks of wheat and maize, and the new characters were found to be heritable.

"However, while it is demonstrated by science that changes can appear suddenly, it still has no knowledge about the causes of such variation. From the observations of Blaringhem's experimental fields I formed an impression that bacteria or other microorganisms could play an important role in the changes found in higher plants.

"Abnormal granule stalks (mutants—B. K.-P.) make an impression of the organs changed by a disease, and thus it is quite natural to search for the causes in parasitism by some microscopic organisms. Small and invisible parasites can be passed from generation to generation. Such are green and brown microscopic algae (zoochlorellae, zooxanthellae), which are inherited in animals in which they live, and which pass to the latter the characters diagnostic for their species."

The same thought is expressed by another prominent biologist, Le Dantec, in his very widely known *Philosophie biologique* [1911], in the following manner: "As our knowledge of nature grows we are convinced again and again that associations or symbioses among different species are widely distributed. It is very possible that numerous plants have symbiosis with microbes, which could be too small to be easily detected under a microscope. Furthermore, it is very possible that a sudden disappearance of one of such microbes, or, on the contrary, its introduction to an organism could result in a large, hereditary variation. According to this hypothesis, a mutation would represent not a sudden appearance of a species but formation or destruction of a new association between two or several species, which existed earlier."

This consideration corresponds well, on the one hand, to the words of De Vries himself: "the cause of progressive mutation lies in the formation (in which way?—B. K.-P.) of new kinds of hereditary primordia"; on the other hand, with certain specific and doubtless facts, e.g., the similarity of a number of classical mutations (e.g., variegated leaf mutations) with known bacterial "diseases"; transmission of some mutations only by females, just like many bacterial and fungal symbioses; etc. It is relevant to note here the fact that bacterial symbionts (in Aleurodidae, above) appear to cause duplication of the chromosomal number in host cells, i.e., are the cause of that "enigmatic" phenomenon that undoubtedly lies at the root of gigantism mutations (cf. *Oenothera gigas*).

Fungi from the groups Entomophoraceae and Isarieae parasitize insects. After they kill their hosts, they emerge and develop spores.

Schaudinn [1904] demonstrated that a certain, probably entomophoracean, fungus is a constant and beneficial symbiont of mosquitoes. Their esophagus has special saclike reservoirs—symbioorgans, formed by the presence of this fungus and usually containing its numerous offshoots (conidia). The fungus is undoubtedly beneficial to the digestion of mosquitoes.[32]

Portier published a series of works, starting in 1911, that demonstrated that isarian fungi are constant and necessary symbionts of Lepidoptera. Mycelium is not only located in the gut but penetrates all body parts of the insect, its muscles, its nervous tissue, and has a beneficial influence on its life. Food particles in the gut may be completely surrounded by mycelium. "The way of feeding in some Lepidoptera is completely dependent on enzymatic characteristics of their symbiotic fungus."

In such cases one might say it is not activity of the fungus that causes the insect's death, but it is rather the insect's death that causes the fungus to emerge and move to its reproductive stage outside.

The stem of common elderberry [*Sambucus nigra*], in its bark, outside from vascular bundles, especially around the core, possesses special sacs—vertical cords, which become brown when they dry out, and are filled by turbid, granular, and viscous mass when alive. These cords develop from single cells by division. These structures were well studied in detail by Dippel, de Bary, and others. Dippel considered them to be lactiferous glands, and de Bary thought that they "store secretions" of "enigmatic significance."

However, the first-class microscopist Oudemans [1872] suggests that these cords, so similar to lactiferous glands, are not a tissue of an organ belonging to the elderberry, but mycelium of a fungus from the genus *Rhizomorpha*. That fungal hyphae are clearly located, relative to vascular bundles, in the same place as lactiferous glands is remarkable.[33]

Another similar example is taken from the well-known zoology textbook by Hertwig, translated into Russian. An organism, *Cytorreictes,* is placed by some researchers among lower organisms and is believed to be a "cause of infection," while others "regard such corpuscles as degenerating blood corpuscles, which are found in normal blood as well."[34]

It is relevant to recall here that the fungal partner of heather is indistinguishable from the parasitic fungus *Phoma,* and that the partner of darnel might also belong to Ustilaginales.

The facts similar to those mentioned above force one to reconsider innumerable cases of "parasitism," "parasitic infections," "mycoses," "bacterioses," etc.

We might possess a great number of examples where complex organisms can be separated into their symbiotic components, examples that reveal the symbiotic nature of various tissues and organs—*but we do not understand them properly.*[35]

Many "pest" fungi are as damaging for their hosts as symbiotic fungi of mosquitoes and Lepidoptera for their cohabitants. The cases when a fungus inevitably drives its partner to death cannot contradict this statement. It is enough to recall various manifestations of autonomy and selfishness of "true" tissues and organs; macrophages, which destroy, to use Metschnikoff's expression, "noble" human elements in old age; osteoclasts, which destroy bone matter; the pituitary gland, with its "selfish" development that leads the organism to the severe disease of acromegalia, etc.

"All physiologists agree that organisms consist of a multitude of independent parts, to a significant degree independent from each other." "Every organ," says Claude Bernard, "is blessed by its own life, its autonomy; it can develop and reproduce itself independent from adjacent tissues." "Every epithelial or muscle cell leads a life of a parasite relative to the rest of the body." And so on. "Many facts support this view of the independent life of each minute element of the body (Darwin [*The Variation of Animals and Plants under Domestication,* ch. 27, p. 365]).

Already, Darwin possessed a significant number of such examples where tissues and organs demonstrated a maximum of autonomy and egoism in different situations. It is enough here to mention general properties of blood corpuscles, leucocytes, and some phenomena characteristic of aging.

Leucocytes, both in structure and activity, are remarkably similar to protozoa such as amoebae. This similarity refers also to their way of feeding, voluntary movement, type of irritability, etc. Leucocytes also remain alive for a long time after their isolation from the host organism.

In the aging organism, some of its elements begin to destroy, to devour others. Some squeeze nerve cells and feed on their contents (Metschnikoff's neuronophages), and, as a result, "the noble nervous tissue is topographically replaced by a very different one, which is physiologically unable to substitute for it" (Metschnikoff [1909]). This results in damage to the nervous system, weakening memory, etc. Other elements damage muscle tissue.

Osteoclasts attack the bone matter, extract calcium and transfer it to other places and other tissues. Because of this, bones become fragile, while calcareous deposits in the blood vessel walls make them less elastic and durable. And so on . . .

The constituent parts of an organism are engaged in the murder of other parts while remaining healthy and viable themselves.

"In plants, we also find the phenomenon of autoparasitism, when one organ lives at the expense of another one in the same individual. In my opinion, it would be a big mistake to see in this phenomenon anything essentially different from real parasitism, or alloparasitism (when another individual serves as a host)"—this was written by the most prominent botanist, Goebel [1923].

As we know, Roux [1881] wrote a treatise on the struggle between the parts of an organism, and Weismann [1918] accepted not only the struggle for existence between tissues, but also "natural tissue selection." Thus, the significance of the organism's parts as, to a great degree, independent and self-sufficient units of life already had to be accepted long ago, when nobody had yet heard about new methods of investigating living objects.

The most recent method, the method of artificial tissue cultures, strengthens the earlier impression even further. Starting with the experiments of Harrison [1907] and Lewis [and Lewis] [1911] and ending with the famous tour de force of Carrel [1912], it was, and still is, possible to grow and reproduce in pure cultures all kinds of tissues, placing their little fragments into corresponding nutrient media. And these little fragments live indefinitely and quite independently, and undergo remarkable change.

Organs exhibit independence, autonomy, and egoism in experimental conditions, as well as in tissues. Those who wish to find some information on this issue can find it even in popular Russian literature, e.g., in Schmidt's books *The Mysteries of Life* [1920] and *The Force of Life* [1922].

"Life of a complex organism,"—wrote this naturalist—"appears to be composed from the life of its separate organs, independent, autonomous, and capable of existing individually outside of the organism. The life of an organism is a sort of a mosaic process, united only by interaction of its parts."

Therefore, (1) great similarity between confirmed symbioorgans and symbiotissues to other organs and tissues, for which symbiogenesis has not yet been proved; (2) existence of such cases when science cannot decide whether it deals, e.g., with a plant tissue or a parasitic fungus; (3) a great autonomy and egoism of "regular" tissues and organs; as well as (4) an unexpectedly rapid growth of the number of confirmed cases of the symbiogenetic origin of tissues and organs—all this leads to the conviction that the majority of, if not all, tissues and organs are a result of symbiogenesis, are symbioorgans and symbiotissues.

One believes that a direct and experimental general proof of symbiogenesis is only a question of time.

IV

The Philosophy of Symbiogenesis

Such facts and specific considerations as those listed above are absolutely, exceptionally interesting by themselves. However, their major importance for science lies, of course, in those general conclusions that can, and should be, made on their basis.

It is easy to see that such generalizations are decidedly different from traditional foundations of biology. They bring along a radical reform of our ideas about the constitution of living organisms, and about its origin.

I will list here only some of the conclusions forming the *credo* of symbiogenetic theory.

1. The [Nucleated] Cell

One of the most important heirlooms passed to the twentieth century by the nineteenth is so-called cell theory. According to this theory, already for three-quarters of a century biology has been convinced that "everything alive is made of cells" (Schwann and Schleiden: "omne vivum e cellulis"); that cells are those architectural elements that comprise organisms; that a cell is "the elementary living machine" or "the most elementary hereditary super-machine" (Reinke, A. Meyer); by a "machine" here is meant "such a collection of bodies, in which these parts make one whole, i.e., are organs serving to achieve a certain goal," a collection, where each body plays a certain role in

---

The major work on the "philosophy" of symbiogenesis remains the paper by Famintsyn [1907].

the interests of the whole; that cell's organelles are a result of differentiation of the homogeneous body of the cell, which took place due to the drive toward labor division.

Recently, cell theory was subjected to severe criticism (by Sinitsin [1923] and others), but this is not enough. The time came to propose, against all these statements, the diametrically opposing ones.

If under "cell" we mean those "elements" of a quite certain structure, a collection of which makes up the majority of organisms, then we admit that there exists also a great number of noncellular organisms, which lack cellular structure (cytodes). Indeed, not every living being is made of cells. [If we recognize the bacteria unit as a cell, which modern science does, Kozo-Polyansky's support of Schwann and Schleiden's "omni cellula" concept is correct.]

A cell also is not an elementary organism because, first, there are organisms with much simpler, more elementary structure (cytodes); second, a cell itself can be separated into a number of more elementary, heterogeneous bioblastic organisms (biotes, in the terminology of Bolaffio [1922]).

Not only is a cell not an elementary machine—it is not a machine at all.

A cell is a system, i.e., a collection of autonomous, self-sufficient bodies that only to a certain degree have lost their individuality. Each exists in its own interests and represents special units of life also outside of the collection in which they are partners.

At the same time, [many of] the cell's organelles are not a product of differentiation but instead are the result of compositeness, joining, and incorporation from outside of the units of life. Prior to incorporation they were autonomous and existed independently.

No drive toward division of labor led to the formation of these organelles, but the incorporation of certain partners into the system ensured the division of labor.

2. Organisms as Consortia

A multicellular organism, divided into tissues and organs, is usually regarded as a colony of cells, i.e., a collection of elements homogeneous in their origin, which only with time acquired more or less different structures, according to their adaptation to various functions.

A multicellular organism is a result of "integration and differentiation" ([Herbert] Spencer), a result of *accumulation* of homogeneous elements and their development along different directions, depending on their function.

Even before cell theory was proposed, biologists accepted the concept of a multicellular organism as a sort of machine. [Many still do. We editors side with Kozo-Polyansky.—*Eds.*]

We, however, now think that there is no basis for considering a multicellular organism as a colony; it is a consortium, a collection of initially heterogeneous units. It was not the difference in function that led to the division of an initially uniform body into its numerous parts, but the coming together of originally heterogeneous units to a complex that could ensure the division of labor.

The composite organism, with its tissues and organs, clearly is not a machine. It is a system of heterogeneous, self-sufficient, more elementary, bioblastic, single-celled and multicellular organisms. It is a federation of organisms.

From the theoretical viewpoint, there is great importance in answering the question of whether an organism is a machine or a system. "There are no machines in inorganic nature," a machine is always a product of an organism, e.g., real machines are products of the "highest degree of organization, i.e., human beings." A machine cannot originate by itself, it has to be made. And so on.

Therefore, if an organism is considered to be a machine, for many it leads to drawing a distinct boundary between organic and inorganic nature, a dualistic worldview, and a necessity to allow for certain "sources" that "made" those machines.

On the contrary, combinations of bodies in inorganic nature can be treated either as aggregates or as systems, and one can think about their origin without involving any supernatural sources.

To argue the concept of an organism as a system appears to be especially valuable, because it is exactly this concept that harmonizes with a monistic and mechanistic worldview.

3. Inherited Variation

Until recently, it was accepted in experimental biology that hereditary change occurs through recombination and mutation. Now we can see that it is symbiogenesis that forms the basis for this process.

Within symbiogenesis, we can distinguish the following stages:

1. Synthesis of two or more heterogeneous organisms into a composite organism, characterized by a new form and a more complex structure

than has each of its partners taken separately. [And two or more genomes to produce a "symbioma" (Guerrero and Margulis 2010)]

2. Interaction of the components of such an organism-system, which determines: (a) movement of these components within a system; (b) extinction of some components; (c) modification of their features; all this, together and separately, results in changes of the characteristics of the entire system.

Natural synthesis of organisms into symbiotic systems can take place between entire organisms, or between any of their parts, or through special primordia.

We have examples of synthesis through adjoining (juxtaposition), penetration, and fusion. In the first case, symbionts are only in close proximity, associated outside of their cells (intercellular symbiosis, etc.); in the second case, one symbiont lives inside a cell (or cells) of another (intracellular symbiosis); in the third case, cells of symbionts fuse, and this fusion may either involve only cytoplasm or extend also to nuclei [the symbioma (Guerrero and Belanguer 2010)].

In the majority of cases one can determine a certain point in time for such synthesis. In some cases probably a constant introduction of symbionts with food takes place during a lifetime; following Portier [1918], feeding has to be regarded not only as intake of nutrients for an organism, but also as intake of beings, which become the organism's partners and join it as its structural ingredients. [Metabolic level of integration or syntrophy (Margulis et al. 2006)]

The sexual act is nothing other than an especially common method for synthesis of a different organism through fusion of special primordia (reproductive cells); that it is nothing but a specific case of symbiogenesis is easy to see. [Margulis 1993, see fig. 24 therein.]

Also, mutations, the hereditary variation in Darwin's sense, "which is triggered by environment but is more dependent on the inner nature of organisms," are probably a result of interaction between a system's components or a struggle among them.

Many authors note that the origin of new forms in the organic world, as it is commonly portrayed by science, is very different from the origin of new forms in the inorganic world.

The statement of Leibnitz, "Natura non operatur per saltum," lost its significance outside of biology long ago.

"In the elementary natural phenomena we see entirely sudden changes: two gases interact to create a liquid; two liquids, a solid residue; sulfur and iron produce crystals, which resemble neither sulfur nor iron," etc., etc. Production of new forms happens in leaps, in jumps, and the new always is a result of integration or disintegration of other bodies.

At the same time, the biology of the nineteenth century has been firmly rooted in an opinion that "Natura non facit saltus" (Linnaeus), and the origin of new bodies is accepted without involving phenomena of synthesis.

At the threshold of the twentieth century, partial reconsiderations of these opinions have been made. The existence of leaps, such as mutations or saltations, has been admitted, as well as a possibility of the origin of new bodies as a result of synthesis or disintegration (Lotsy's combination theory and related concepts).

However, mutations are such insignificant leaps that, as observed (by Plate[1] and others), the same examples of the origin of new forms, which some understand as leaps (De Vries), to others appear to be examples of gradual change (Darwin).

Those researchers who accept the role of synthesis in the process of the origin of forms only in the case of interbreeding are forced to introduce another, additional explanation for organisms that lack a sexual process. They must explain large innovations in structure, because the creative scale of hybridization is not too high, only very similar forms interbreed, and the new differs very little from the old.

The study of symbiogenetic phenomena forces one to accept that, first, the origin of new forms through synthesis takes place in organic as well as in inorganic nature, and at this point, instead of previously assumed principal differences one is led to accept a remarkable unity of phenomena.

Second, the statement "Natura non facit saltus" now loses its significance also in the sphere of phenomena related to the origin of forms in organic nature. Here it is also not more applicable than in physical phenomena and chemical reactions: formation from two or more organisms of a composite third must be interpreted as a "leap." [And verified in the fossil record as "punctuated equilibrium" (Guerrero and Margulis 2010)]

Currently, the existence of three categories of hereditary variation can be considered as demonstrated in nature:

1. Combinations, i.e., new entities formed through association of heterogeneous reproductive cells during interbreeding.

2. Mutations, i.e., new entities formed besides interbreeding, due to internal "enigmatic" causes.

3. Fusions, i.e., new composite entities formed due to the association of heterogeneous organisms besides that mediated by gametes, i.e., through joining of certain specialized primordia, which combine in a certain way.

The basis for all three categories is symbiogenesis, and at the same time they [combinations, mutations, and fusions] assure not only leaps and gaps [missing links] but also continuity. Symbiogenesis, by fusion, obviously provides the largest scale discontinuities. In those cases where "saltations" are so large relative to e.g., mutations we have every right to call them as such,

I will dwell in more detail on the analysis of "leaps" in organic nature in my other book, *Dialectics in Biology*.[2]

4. Heredity

We have seen in many examples that every partner of a symbiotic system reproduces separately. Each organelle, as well as each tissue and organ, the existence of which is due to symbiogenesis, all lower units comprising organisms, reproduce separately. They often form specific primordia, which serve for hereditary transmission.

Such primordia can be multicellular or unicellular [eukaryotic], or they can be of bioblast [prokaryotic] nature, often at the limit of visibility.

Such primordia, in most cases, move actively or passively in some way, concentrate in certain locations, and form more or less complex aggregates, or systems of primordia. These aggregates serve for the reproduction of entire symbiotic systems.

Examples of highly specialized sets of hereditary primordia are reproductive cells.

Development of an organism-system from the egg is a process, which merges growth, reproduction, and interaction of all partners in this system, represented in the egg as primordia.

Primordia for very many features of complex organisms are undoubtedly located outside of the nucleus of reproductive cells, outside of mitochondria or similar bodies.

The entire reproductive cell is a collection of hereditary primordia.

Therefore, the material on which the theory of nomogenesis operates gives a possibility for direct proof of the hereditary theory of "congenesis" [pangenesis] proposed by Darwin.

This "temporary hypothesis," as the author modestly called it, completely rejected by science and named "unfortunate" and "fantastic," received a brilliant confirmation not only in its first part—the assumption that hereditary primordia (gemmules) are produced by an organism's parts, but also its second part, which was considered by pangeneticists themselves to be difficult to prove—the assumption that hereditary primordia are transferred to reproductive organs.

Finally, Harvey's statement, firmly professed by all biologists until recently, "Omne vivum ex ovo," if by "egg" we mean "zygote,"* cannot be retained without reservations.

Not even mentioning that many organisms lack not only a sexual process but even a cell structure, one has to take into account, first, the hereditary transmission of tissues and organs (or their ability to form) besides reproductive cells and zygote (e.g., roots in *Erica,* heather); second, mandatory development of an individual from an aggregate of several, at minimum two, separate cells (e.g., in lichens); third, lifelong potential to accept new partners into the system through new infection in other ways (e.g., in orchids).

5. Natural Selection and the Struggle for Existence

If one believes that an organism is a machine, then one has to accept that the struggle for existence extends from social units to the "indivisible," i.e., the entire organism. It seems absurd to talk about the struggle for existence among parts of a machine, which were "made" with an aim to preserve a certain form, take a certain place, and play a certain role—to be *parts of a whole.*

However, as we already noted, the facts accumulated long ago that, willy-nilly, forced the biologists to accept the reality of struggle among organs, struggle among tissues, and therefore, a possibility, for example, of "tissue selection" (Weismann). Such facts, however, did not get due attention, and the very expression "struggle among the parts" (Roux [1881]) was understood by the majority of researchers as a metaphor of a kind.

*Product of fusion of male and female reproductive cells.

Weismann [1918], however, one of the few who accepted the importance of struggle and selection inside an organism, emphasized that "applying the principle of struggle and selection to the living units of all levels, down to and *including cells* constitutes the center of gravity of [his] views," the "center of gravity" of neo-Darwinism, and a large step forward in the understanding of the mechanism of evolution.

Now science is able to make two other significantly important steps forward.

As it rejects understanding of the "machine" nature of organisms, and accepts them as systems of self-sufficient, heterogeneous beings, the theory of symbiogenesis becomes able not only to accept the naturalness and necessity of *internal* struggle and internal selection, but also to understand how these phenomena take place inside an imaginary "indivisible," and to see nothing paradoxical or unnatural in such phenomena.

When one recognizes the [eukaryotic] cell as the sum of elementary organisms, as [prokaryotic] consortia, one admits the reality of *intracellular* struggle and *intracellular* selection, i.e., struggle and selection between partners, between nuclear components, chromosomes, and other "organelles."

It is time to transfer the principle of struggle and selection *even further* than it was thought by Weismann himself.

Thus, a new understanding of the constitution of organisms makes one confident that the meaning of the principle of struggle and selection is even more wide and all-embracing than was assumed by Darwin and other selectionists. "Der Wiederspruch ist das Fortleitende" [Contradiction leads the way forward] (Hegel).[3] This thesis of dialectics is acceptable now down to mitochondria and chromioles [chromosomes]!

In this principle of internal, endosomatic, and intracellular struggle, biology can possibly find a key to understanding the causes of mutations and other hereditary phenomena.

6. Symbiogenesis as Evidence for Evolution

Until recently, it was widely accepted that the reality of evolutionary process can be confirmed only indirectly, and that this process cannot be demonstrated through a direct, and especially experimental, proof. Experimentally, we observe only the origin of new forms [variation], but not evolution, which is the origin of "more perfect" forms.

Therefore, many researchers assumed, and continue to assume, that the evolutionary doctrine belongs to the domain of faith and is a belief that cannot be proved.

However, such examples as the origin of lichens from association of algae and fungi are examples that allow direct observational and experimental testing of the origin of complex organisms. Composite organisms persist in the struggle for existence relative to the rather simple and less adapted organisms from which they evolved.

These examples, without a doubt, provide a direct and experimental proof for the reality of evolutionary process (Weismann's argument).

Therefore, the theory of symbiogenesis possesses that proof of the reality of evolution, which was considered a *pium desiderium* ["holy longing"] of biology.

This is not all: we have to accept that in cases where a direct and experimental proof for evolution is possible we invariably encounter the phenomenon of symbiogenesis. It is obvious that this interpretation of complex organisms as systems of elementary units provides a firm foundation for the reality of the evolutionary process.

7. "Missing Links"

Almost all evolutionists are involved in the search for "intermediate," "transitional" types and forms, since the existence of such forms is believed to be postulated by the theory of evolution. The absence of such intermediate links in desired cases and the rarity of their discovery in general led many biologists, beginning with Darwin himself, either to express their bewilderment or to suggest that our knowledge is "incomplete," etc.

From the viewpoint of concepts developed here, the search for transitional forms and structure in many instances can indeed be considered fruitless and aimless. No transition existed between the composite obtained as a result of integration of heterogeneous components, and these components themselves. Some organs were formed due to invasion of corresponding symbionts, while closely related species may have no slightest hint of such symbioorgans.

8. Genealogical Relations: [Anastomoses of Family Tree Branches]

According to Darwin, the best manner in which to present relatedness among living organisms is a tree with multiple divergent branches, a "family tree" similar in shape to real trees.

This opinion is accepted by the overwhelming majority of evolutionists. According to some, however ([Leo] Berg [1922]), genealogical relationships should be presented in the shape of a rye field—as numerous parallel, similar phyletic rows. Elsewhere,* I attempted to show that this fantasy has no validity.

Without a doubt, a "genealogical tree," as an expression of these phenomena, should be preferred. Everything described here demonstrates that in this case we also have to make a radical [topological] correction.

A genealogical tree of organic nature does not follow the shape of a real tree. The genealogy of organisms is expressed not only in divergence of lineages but also in fusion, or joining; two or more branches fuse and give a merged lineage [anastomosis]. This branch, in its turn, can fuse or merge with other branches to give a new branch of higher order, etc.

One has to keep in mind, however, that this fusion, i.e., the association of several types of organisms to make a composite type (fungus + alga = lichen) has nothing to do with convergence in the sense of Danilevsky, Berg, etc., where this term designates full *similarity* among several organisms belonging to different genetic lineages.

9. Issues in Evolution

It has been difficult to provide natural explanations for the origin of some specific organs. From ancient times to Bergson [1914], mystics and metaphysicists, for example, liked to speculate about the eye. Even Darwin's supporters considered the eye to be a sort of stumbling stone for the materialistic dialectic. There are curious documents about this in Darwin's *Life and Letters.* Following the advice of his friends (Charles Lyell, Asa Gray, and others), Darwin removed from the galleys of the first edition of *Origin of Species* the place where he mentioned genesis of the eye.

---

*See my book *The Last Word of Anti-Darwinism*, 1923 ("Burevestnik").

Given a certain similarity between the highly organized eye and the cephalopod luminescent organ, in connection with the theory of symbiogenesis in general, one can't help but come to the possibility of a completely novel approach to understanding the genesis of a complex vision device. This approach—the approach based on the ideas of symbiogenesis—is completely devoid of mysticism, which makes it especially valuable in this case favored by idealists, and in similar cases.

There also has been quite a lot of tension about the phenomena of convergence in organ structure. The fact, e.g., that such different organisms as cephalopods and vertebrates possess similar complex organs of vision was presented by Mivart[4] and others as a tool to discredit evolutionism and Darwinism, as evidence for preordination, etc. Romanes agreed that if full convergence would be demonstrated here, it would be "deadly." From the perspective of symbiogenetic theory, the existence of similar organs in different sections of the organic world has nothing mystical, enigmatic, or supporting of metaphysics or teleology. If the organ in question is a symbioorgan, if a given feature results from synthesis, it can appear at any point where a corresponding symbiotic association takes place.

"Let us imagine,"—writes Merezhkovsky [1909a] "a palm peacefully growing by a brook, and a lion, hidden in the bushes, with flexed muscles, with bloodshot eyes, ready to pounce on an antelope. Only the theory of symbiosis can reach to the very bottom of the mystery contained in this picture, only this theory can shed light on the fundamental cause of an enormous difference between such phenomena of life as a lion and a palm.

"A palm is peaceful and passive exactly because it is a symbiotic system; because it contains an entire crowd of tiny green toilers, the chloroplasts. They work and they feed it. And a lion feeds by itself.

"But let us imagine that a chloroplast is placed in every one of a lion's cells, and I have no doubt that this lion will then calmly lie next to the palm, and the only other thing it might need would be a little water with mineral salts in it."

It is easy to see that the theory of symbiogenesis allows addressing, from a new perspective, many important, specific issues of the evolution of organic nature, up to the issue of how it diverged into two ways: the animal kingdom and the plant kingdom.

10. A Program for Biology

The merits of various new theories are expressed in different ways. Their ability to embrace more facts, to incorporate these or other facts better than was done by previous theories, their ability to explain facts that could not be explained before—all this points at the viability and scientific value of our theory.

However, another criterion can be also suggested.

Truly new and powerful ideas always possess a gift to open new areas of research, to point at new ways of study; in a word, they present new programs of action for science, they facilitate innovation.

The theory of symbiogenesis, without a doubt, presents a new and broad program for biology. The motto of this program is: particentrism instead of toticentrism; *shift the focus of attention in the study of living bodies from the whole to its parts.*

To find, for each complex organism, the components of which it is made; to study the structure, way of life, systematics, distribution of each of those components separately, by isolation from consortia as well as by observation inside the system; to perform analysis and synthesis of the studied consortial systems—these are the current tasks of biology generated by the theory of symbiogenesis.

Only after resolving this issue will we be able to reach an understanding of structure and function of systems, of complex organisms taken as a whole.

One can only partially predict what incredible discoveries await biology on this new path. It is even more difficult to tell how this new direction of biological science would reflect on related sciences, although there is no doubt that, similar to its previous successes, this one will also evoke a response far beyond the limits of biology.

One can think, for example, that in sociology, the organic theory of society, in its classical Spencer–Lilienfeld version, will completely lose its significance;[*5] and that psychology will finally solve the issue of the causes

---

*The version of Gumplowicz seems to have no objections from our theory's perspective.

of "split personality," of the natural basis for that condition, familiar to many, which is reflected in the famous lines:

> *Zwei Seelen wohnen, ach! in meinen Brust*
> *Die eine will sich von der andern trennen* . . . [6]

Understanding of nature leads to its submission to man. The practical consequences of the theory of symbiogenesis are already seen both in medicine and in agriculture, i.e., in both areas of applied biology. Those will have to be discussed elsewhere.

V

History of Symbiogenesis Theory

The beginnings of the basic theory of symbiogenesis, as well as many other most important ideas that inspired modern thought, can be found in some of the ancient Greek philosophers.

Empedocles (492–432 BC) taught in an especially vivid, and at the same time crude and fantastic, way about the origin of complex and perfect organisms through associations of the parts that existed separately. First, various independent organs appeared: "heads without ears, eyes without head, hands without bodies." Then, "under the influence of love, these parts started to associate without order." But "everything that did not fit the natural combinations, died" or fell apart into the initial elements.

"The doctrine of Empedocles is based on combination of heterogeneous forms with each other" (Lange [1881]). The essence of this doctrine is the same as in our theory, but the former was a pure fantasy, since at this time no facts existed, or could exist, to support it. Besides, the very way of formulating the idea is unacceptable from the modern viewpoint.

The purely speculative stage in the life of the idea of symbiogenesis starts and ends with the Greeks.

After Empedocles, we have to jump directly to the second half of the "Age of Natural Science." All through this period, Aristotle's opinion reigned unchallenged that the "whole exists before parts, the parts grow on the whole and from the whole, there is no other way of their origin" (Paulsen [1904]). "The tiniest hair does not appear anywhere else but on the body to which it belongs, whatever shakeup we apply to the atoms." "To believe (as follows from Empedocles) that the hair in a lion's mane, appearing separately and in hundreds of thousands flying through the air, suddenly would gather on certain single skin" seemed an incredible absurdity. Currently, however, we have such an example as well, of free-living hair, although not a lion's but a dragonfly's (see chap. III, 1).

Some ([Leo] Berg) think that Oken, around 1805, presented a theory identical or close to our theory. He taught that an animal body is a collection of infusoria, etc.

It would be, however, unfair to classify Oken as a precursor of the theory of symbiogenesis. As can be seen from his *Lehrbuch der Naturphilosophie,* his "infusoria" are equivalent to the "cells" of cell theory, nascent at this time, i.e., Oken considered a complex organism—translating specific terminology of this author to modern scientific language—as a colony of cells. His view is a trivial view of biology of the nineteenth century. The main point in the theory of symbiogenesis—the principle of synthesis of *heterogeneous* organisms in a system-consortium—was quite alien to Oken's theory.

The new history of this theory begins with Charles Darwin. He can be considered the real progenitor of the new principle of biology. This should be accepted not only because, as it seems to me, this great evolutionist formulated the idea of the theory of symbiogenesis very clearly and in a modern scientific manner, but especially because he came to it through an inductive way and collected a large number of indirect proofs in its support.

Fairness requires, however, mentioning here the name of Brücke [1862], the author of the concept of cells as elementary organisms. In his famous article "Die Elementarorganismen," published in *Sitzungsber. Wien. Acad.* (44, 1862), we find such prophetic lines: "I call the cell an elementary organism in the same sense as in chemistry we call elementary such bodies which have not yet been taken apart. However, just as the indivisibility of chemical elements has not yet been proven, it would be unfair to deny a possibility that cells themselves, in their turn, are composed of other even smaller organisms, which have the same relationship to cells as these latter have to the entire organism. Until recently, however, we had no proofs to accept this viewpoint."

Already in 1868, in chapter 27 ["Provisional Hypothesis on Pangenesis"] of his book *The Variation of Animals and Plants under Domestication,* Darwin presented opinions of the following kind.

The cell is not an element but, in its turn, consists of a large number of elements similar to bacteria.

"An organic being is a microcosm—a little universe, formed of a host of self-propagating organisms, inconceivably minute and numerous as the stars in heaven." In other words, a complex organism is a system of "independent organic units."

When a complex organism reproduces, each of its components, each "living unit" comprising it, is reproduced. Sexual elements include primordia ("gemmules") of all components of the organism. Therefore, an egg cell, for example, is a system of offshoots from all the units, which composed the organism that produced it. And so on.

In Darwin's opinion, only when we accept these views will we be able to explain the paradoxical independence of tissues and organs from each other, the existence of graft hybrids, to understand "how a certain limb can be restored right along the amputation line," how the same organism can originate through such different processes as budding and seed reproduction, etc.

One can say that Darwin was led to acceptance of the principle of symbiogenesis, to the understanding that the organism is a sum of more elementary organisms, by his study of phenomena of heredity and physiological independence of tissues and organs.

By assigning great significance to the phenomena observed during explantation and transplantation, Darwin in fact stated a possibility of direct and experimental proof of "microcosmic," consortial nature of complex organisms.

It appears that Darwin did not even suspect that similar proof was, in fact, established in the same year for an entire division of plants. We are talking about decomposition of various lichens into more elementary organisms, and further artificial synthesis of the latter.

However, we must keep in mind that, accepting not only a complex organism but even every single cell as a "microcosm," Darwin did not specify a manner of combination of the elements into a system, the way by which this microcosm is created.

The fate of Darwin's pangenesis theory is well known and instructive. The theory was rejected and even ridiculed by the majority of scientists. Even Wallace called it "unfortunate" and "fantastic." Few attempted to defend it, evidently without understanding its essence (De Vries, who wrote a book,

[*Intracellular*] *Pangenesis* [1889]). A remarkable [Darwin's] concluding statement on pangenesis, "An organic being is a microcosm," etc., was seen as a metaphor; De Vries applied it to the chromosomal theory of heredity (epigraph of his book mentioned above), and Hertwig to cell theory (epigraph to *Cell and Tissues* [*Die Zelle und die Gewebe* (1893–1898)]).

Darwin himself, however, understood the meaning of this great idea and did not reject his hypothesis. In his autobiography we can find the famous words: ". . . my well-abused hypothesis of Pangenesis. An unverified hypothesis is of little or no value;* but if anyone should hereafter be led to make observations by which some such hypothesis could be established, I shall have done good service, as an astonishing number of isolated facts can be thus connected together and rendered intelligible."

Today, the theory of pangenesis is also usually rejected[†] and placed among the unsupported concepts of heredity, i.e., doctrines that are of low importance. Already for Darwin himself, however, this concept was much less speculative than hypotheses of Weismann, De Vries, Naegeli, etc. Since that time, direct confirmations of this concept have been accumulated. Also overlooked is that "none of the modern theories of heredity, which appropriated from the theory of pangenesis its idea of representative units, added anything significant to the explanations provided by this theory. Without Darwin's original theory they would not have the foundation on which they elaborated" (Delage [and Goldsmith] [1916]). Finally, using the words of De Vries, the theory of mutations, which is so glorified today, is "a child of pangenesis."

"In all problems of biology Darwin has to be credited by being everywhere a major great initiator" (Delage [and Goldsmith] [1916]).

Reinke [1873] has the honor of being first to admit that lichens have a great theoretical value; it is lichens that provide a direct and experimental proof of the existence of such multicellular organisms, which are sums of other, less complex multicellular organisms.

Thus, Reinke can be called the first experimental symbiogeneticist.

---

*A direct proof is probably meant.
†See, e.g., Filipchenko [1917, 1924], *Heredity*.

The next most important moment in the history of our theory is the discovery of the true nature of chloroplasts in animals, i.e., of zoochlorellae and zooxanthellae, by Geza Entz [1881] and Brandt [1881–1883].

This discovery established, by direct observation and experimentally, that an independent organism can be an organelle of a cell. Thus began the artificial analysis and synthesis of a cell, the study of its separate "gemmules," which probably Darwin himself did not dream of.

Therefore by the second half of the nineteenth century the main principle of the theory of symbiogenesis had been not only derived by induction, but its general principles were clearly formulated with respect to cells and multicellular organisms. Furthermore, the possibility of its direct, experimental confirmation also has been demonstrated.

It is not surprising therefore that in the nineteenth century already we can list a number of researchers who to some degree accepted the symbiogenetic viewpoint. Among them are well-known names such as Altmann, N. Bernard, Béchamp, Boveri, Wigand, Le Dantec, Merezhkovsky,* E. Metschnikoff, Famintsyn, Schimper, and others.[†]

However, for the principle of symbiogenesis to cease being a private opinion of certain individuals who are able to overcome the shackles of tradition, for it to become a property of wide scientific circles and to attract the attention of thinking people in general, it was necessary to collect a great number of examples, which would confirm this principle as correct and universal. This has been done already in the second decade of the twentieth century by the works of Buchner, Lyubimenko, Peklo, Pierantoni, Portier, Rayner, Reichenow, Šulc, and other representatives of biological science in the most recent times.

---

*The term "symbiogenesis" itself belongs to Merezhkovsky [1909a] and was suggested in his book, *The Theory of Two Plasms.*

†Merezhkovsky [1905, 1909a, 1910] and, especially confidently, Famintsyn [1907, 1912], have been until recently the only supporters of the theory of symbiogenesis in Russia. Most recently, support of the symbiogenetic viewpoint has been also voiced by Lyubimenko [1916, 1918] and Karpov [1922] (both regarding the genesis of a cell) as well as Keller [1924].

Even today, for many, of course, the theory of symbiogenesis would seem paradoxical—moreover, improbable. But "when it was first said that the sun stood still and the world turned round, the common sense of mankind declared the doctrine false."*

And yet it does move![1]

---

*Darwin, *The Origin of Species,* ch. 4 [passage added to later editions of *The Origin of Species* (1872, p. 134)—*Eds.*].

References to Kozo-Polyansky's Text

Altmann, R. 1890. *Die Elementarorganismen und ihre Beziehungen zu den Zellen.* Leipzig: Veit and Co.

Areschoug, J. E. 1854. *"Spongocladia,* ett nytt algslägte." *Öfversigt af Kongl.* [*Svenska*] *Vetenskaps-Akademiens Förhandlingar* 10: 201–209.

Arnoldi, V. M. 1901. *Vvedenie v izuchenie nizshikh organizmov* (Introduction to the study of lower organisms). Moscow.

Aynaud, M. 1909. *Le globulin des mammifères.* Paris: Steinheil.

Baden, M. L. 1915. "Observations on the germination of the spores of *Coprinus sterquilinus,* Fr." *Annals of Botany* os-29: 135–136.

Balbiani, E. G. 1864. "Sur la constitution du germe dans l'œuf animal avant la fecondation." *Comptes rendus de l'Académie des Sciences de Paris* 58: 584–588, 621–625.

———. 1869–1872. "Mémoire sur la génération des Aphides." *Annales des sciences naturelles—Zoologie et biologie animale* 5, 11: 5–89 (1869), 14, Art. 2 (1870), 15, Art. 1 (1872).

Bambacioni, V. 1920. "Sulle struttura di fibrilli del Němec." *Atti della Reale Accademia dei Lincei, Cl. Fis.* 29a: 62.

Baranetsky, O. 1868. "Independently living gonidia of lichens" [in Russian]. *Trudy Syezda russkikh yestestvoispytatelei i vrachei. Otdel bot.* (Transactions of the First Russian Congress of Naturalists and Physicians. Division of Botany). St. Petersburg, pp. 45–59.

Bary, A. de. 1879. "Die Erscheinung der Symbiose." In *Vortrag auf der Versammlung der Naturforscher und Ärtze in Cassel.* Strassburg: Trubner, pp. 1–30.

Baumgärtel, O. 1920. "Das Problem der Cyanophyceenzelle." *Archiv für Protisten-kunde* 41: 50–141.

Baur, E. 1909. "Das Wesen und die Erblichkeitsverhältnisse der 'Varietates albomarginatae Hort.' von *Pelargonium zonale." Zeitschrift für induktive Abstammungs- und Vererbungslehre* 1: 330–352.

————. 1922. *Einführung in die experimentelle Vererbungslehre*, vol. 6, A. Berlin: Bornträger.

Beer, R. 1909. "On elaioplasts." *Annals of Botany* 23: 63–72.

Beijerinck, M. W. 1890. "Kulturversuche mit Zoochlorellen, Lichenogonidien unde andered niederen Algen." *Botanische Zeitung* 48: 125–139, 741–754, 757–768, 781–785.

Berg, L. 1922. *Teorii evolyutsii* (Theories of evolution). Petrograd: Academia.

Bergson, H. 1914. *Tvorcheskaya evolyutsiya* (Creative evolution). Moscow-St. Petersburg: Russkaya mysl'. (Translation of *L'evolution creative*)

Bernard, N. 1909. "L'évolution dans la symbiose. Les orchidées et leurs champignons commensaux." *Annales des Sciences Naturelles de Paris, Botanique, Paris* 9: 1–196.

————. 1916. *L'evolution des plantes.* Paris: Alcan.

Blochmann, F. 1884. "Über die Metamorphose der Kerne in den Ovarialeiern und über den Beginn der Blastodermbildung bei den Ameisen." *Verhandlungen des naturhistorisch-medicinischen Vereins zu Heidelberg* 3: 243–247.

Bolaffio, C. 1922. *I Bioti: Abbozzo di una nuova teoria della struttura della cellula.* Trieste: Tip. Sociale.

Born, G. 1894. "Die künstliche Vereinigung lebender Theilstücke von Amphibien-Larven." *Jahresbericht der schlesischen Gesellschaft für vaterländische Kultur*, 8 June 1894: 1–13.

Borodin, I. P. 1910. *Kurs anatomii rastenii* (The course of plant anatomy), 4th ed. St. Petersburg.

Boveri, T. 1887. *Zellenstudien I: Die Bildung der Richtungskörper bei* Ascaris megalo-cephala *und* Ascaris lumbricoides. Jena: G. Fischer.

Bower, F. O. 1923. *The Ferns.* Vol. 1: *Analytical Examination of the Criteria of Comparison.* New York: MacMillan and Co.

Brandt, K. [Kozo-Polyansky does not quote an exact work of the 1880s. The following 1881–1883 works by Brandt are found in Buchner (1965)—Eds.]:

————. 1881. "Über das Zusammenleben von Thieren und Algen." *Verhandlungen der Physiologischen Gesellschaft Berlin* 1881–1882: 22–26.

————. 1882. "Über die morphologische und physiologische Bedeutung des Chlorophylls bei Thieren." *Archiv für Anatomie und Physiologie, Abt. Physiol.:* 125–151.

————. 1883a. "Über die morphologische und physiologische Bedeutung des Chlorophylls bei Thieren." *Mitteilungen aus der Zoologischen Station zu Neapel* 4: 191–302.

————. 1883b. "Über die Symbiose von Algen und Thieren." *Archiv für Anatomie und Physiologie:* 445–454.

Brefeld, O. 1872–1912. *Untersuchungen aus dem Gesamtgebiete der Mykologie.* Leipzig: Münster.

Brücke, E. W. 1862. "Die Elementarorganismen." *Sitzungsberichte der Kaiserlichen Akademie der Wissenschaften in Wien. Mathematisch-naturwissenschaftliche Klasse* 44: 381–406.

Buchner, P. 1912. "Studien an intrazellulären Symbionten. I. Die Symbionten der Hempiteren." *Archiv für Protistenkunde* 26: 1–116.

———. 1914. "Sind die Leuchtorgane Pilzorgane?" *Zoologischer Anzeiger* 45: 17–21.

———. 1920. "Zur Kenntnis der Symbiose niederer pflanzlicher Organismen mit Pedikuliden." *Biologisches Zentralblatt* 39: 535–540.

———. 1921. *Tier und Pflanze in Intrazellulärer Symbiose.* Berlin: Bornträger.

———. 1923. "Studien an intrazellulären Symbioten. IV. Die Bakteriensymbiose der Bettwanze." *Archiv für Protistenkunde* 46: 225–263.

Buder, J. 1914. *"Chloronium mirabile."* *Berichte der Deutschen Botanischen Gesellschaft* 31: 80–97.

Burgeff, H. 1909. *Die Wurzelpilze der Orchideen, ihre Kultur und ihr Leben in der Pflanze.* Jena: G. Fischer.

———. 1912. "Über Sexualität, Variabilität und Vererbung bei *Phycomyces nitens.*" *Berichte der Deutschen Botanischen Gesellschaft* 30: 679–685.*

Bütschli, O. 1892. *Untersuchungen über mikroskopische Schäume und das Protoplasma,* vol. 1. Leipzig: W. Engelmann.

Carnoy, J. B. 1884. *La biologie cellulaire.* Lierre: Van In et cie.

Carrel, A. 1912. "On the permanent life of tissues outside of the organism." *Journal of Experimental Medicine* 15: 516–528.

Carter, H. J. 1878. "Parasites of the Spongida." *Annals and Magazine of Natural History,* ser. 5, 2: 157–172.

Cerfontaine, P. 1890. "Recherches sur le système cutané et sur le système musculaire du lombric terrestre." *Archives de Biologie* 10: 327–428.

Chamberlain, C. J. 1921. In *Morphology of Gymnosperms,* by J. M. Coulter and C. J. Chamberlain. Chicago: University of Chicago Press.

Cienkowski, L. 1871. "Über Schwärmerbildung bei Radiolarien." *Archiv für mikroskopische Anatomie* 7: 372–381.

Conn, H. W., and H. J. Conn. 1923. *Bacteriology: A Study of Microörganisms and Their Relation to Human Welfare, Discussing the History of Bacteriology, the Nature of Microörganisms, and Their Significance in Connection with Pathology, Hygiene, Agriculture and the Industries.* Baltimore: Williams and Wilkins.

Cowdry, E. V., and P. K. Olitsky. 1922. "Differences between mitochondria and bacteria." *Journal of Experimental Medicine* 36: 521–533.

Crampton, H. E. 1898. "Coalescence experiments upon the Lepidoptera." In *Biological Lectures Delivered at the Marine Biological Laboratory of Wood's Holl* [sic—Ed.], *Boston, 1896–1897,* pp. 219–230.

Crato, E. 1893. "Morphologische und mikrochemische Untersuchungen über die Physoden." *Botanische Zeitung* 51: 157–195.

*Newer works of the same author in *Flora* 1914, 1915 remained unaccessible for me [B. Kozo-Polyansky].

Cuenot, L. 1898. "Études histophysiologiques sur les oligochetes." *Archives de Biologie* 15: 79–124.

Dangeard, P. A. 1899. "Études sur la cellule. Son évolution, sa structure, son mode de reproduction." *Botaniste* 6: 1–292.

———. 1921. "La structure de la cellule végétale dans ses rapports avec la théorie du chondriome." *Comptes rendus hebdomadaires des séances de l'Académie des Sciences de Paris* 173: 120–123.

Darwin, C. 1868. *The Variation of Animals and Plants under Domestication,* 1st ed. London: John Murray.

Davis, B. M. 1894. "Notes on the life-history of a blue-green motile cell." *Botanical Gazette* 19: 96–102.

Delage, Y. [and M. Goldsmith]. 1916. *Teorii evolyutsii* (Theories of evolution). Petrograd: Izd-vo Popova. (Translation of *Les théories de l'evolution.*)

Derschau, M. von. 1920. "Pflanzliche Plasmastrukturen und ihre Beziehungen zum Zellkern." *Flora* 113: 199–212.

De Vries, H. 1889. *Intrazelluläre Pangenesis.* Jena: G. Fischer.

Drew, A. H. 1920. "Preliminary tests on the homologue of the Golgi apparatus in plants." *Journal of the Royal Microscopic Society* 1920: 295–297.

Dubois, R. 1914. *La vie et la lumiére.* Paris: Alcan.

Elenkin, A. A. 1921. "Lichens as objects of pedagogy and scientific research" [in Russian]. *Ekskursionnoe delo* (Field science), 2–3: 114–178.[*][1]

Enderlein, G. 1921. "Über der geschlechtliche Fortpflanzung der Bakterien." *Beiheft zum botanischen Zentralblatt* 38: 53.

Entz, G. 1881. "Über die Natur der 'Chlorophyllkörperchen' niederer Tiere." *Biologisches Centralblatt* 1: 646–650.

Eriksson, J. 1910. "Über die Mykoplasmatheorie, ihre Geschichte und ihren Tagesstand." *Biologisches Centralblatt* 30: 618–623.

———. 1921. "The mycoplasm theory—Is it dispensable or not?" *Phytopathology* 11: 385–388.

Escherich, K. 1900. "Über das regelmaessige Vorkommen von Sprosspilzen in dem Darmepithel eines Käfers." *Biologisches Centralblatt* 20: 350–358.

Ewart, A. J. 1897. "On the evolution of oxygen from coloured bacteria." *Journal of the Linnean Society of London, Botany* 33: 123–155.[†]

Famintsyn, A. 1907. "The role of symbiosis in the evolution of organisms" [in Russian]. *Mémoires de l'Académie Impériale des Sciences de St.Pétersbourg,* sér. 8, 20: 1–14.

[*] A curious type of a lichen, or something else for which no name yet exists, was recently discovered by Vainio (1921). These are spots on tree leaves (in Manila). They are formed by the mycorrhyzae of two fungi (one of which is not found separately at all), and the cells of one of these fungi are inhabited by an alga (*Cephaleuros virescens*) Both fungi proved to be new for science. They belong to two different genera, *Diplothrix* and *Gonidiomyces.*

[†] Compare: Lyubimenko (1916).

———. 1912. "Die Symbiose als Mittel der Synthese von Organismen." *Berichte der Deutschen Botanischen Gesellschaft* 30: 435–442.

Famintsyn, A., and O. Baranetsky. 1867. "Zur Entwicklungsgeschichte der Gonidien und Zoosporen: Bildung der Flechten." *Mémoires de l'Académie Impériale des Sciences de St.-Pétersbourg,* sér. 8, 11: 1–6.

Fauré-Fremiet, E. 1910. "Étude sur les mitochondries des protozoaires et des cellules sexuelles." *Archives d'anatomie microscopique* 11: 457–648.

Fayod, V. 1891. "Structure du protoplasme vivant." *Revue générale de botanique* 3: 193–228.

Filipchenko, Yu. A. 1917. *Nasledstvennost'* (Heredity). Moscow: Priroda.

———. 1924. *Nasledstvennost'* (Heredity). 2nd ed. Leningrad.

Fischer, A. 1897. *Untersuchungen über den Bau der Cyanophyceen und Bakterien.* Jena: G. Fischer.

Flemming, W. 1882. *Zellsubstanz, Kern und Zelltheilung.* Leipzig: Vogel.

Francé, R. H. 1908. *Das Leben der Pflanze,* 2, III. Stuttgart: Kosmos, Handweiser für Naturfreunde.

Frank, B. 1885. "Über die auf Wurzelsymbiosen beruhende Ernährung gewisser Bäume durch unterirdische Pilze." *Berichte der Deutschen Botanischen Gesellschaft* 3: 128–145.

Freeman, E. M. 1904. "The seed fungus of *Lolium temulentum* L., the darnel." *Philosophical Transactions of the Royal Society of London,* ser. B, 196: 1–27.

Frommann, C. 1880. *Beobachtungen über Struktur und Bewegungserscheinungen des Protoplasmas der Pflanzenzellen.* Jena: G. Fischer.

Fuchs, J. 1911. "Über die Beziehungen von Agariceen und anderen humusbewohnenden Pilzen zur Mycorhizenbildung der Waldbäume." *Bibliotheca Botanica* 76: 1–32.

Fuller, G. D. 1922. "Mycorrhiza of forest trees." *Botanical Gazette* 73: 506.

Geddes, P. 1882. "Further researches on animals containing chlorophyll." *Nature* 25: 303–304.

Geitler, L. 1922. "Zur Cytologie der Blaualgen." *Archiv für Protistenkunde* 45: 413–418.

Giard, A. 1888. "Sur les *Nephromyces,* genre nouveau de champignons parasites du rein des Molgulidées." *Comptes rendus de l'Académie des Sciences de Paris* 106: 1180–1182.

Goebel, K. 1915–1923. *Organographie der Pflanzen.* vol. II, pt. 1, 1915; vol. II, pt. 2, 1918; vol. III, pt. 3, 1923. Jena: G. Fischer.

Graff, L. von. 1891. *Die Organisation der Turbellaria acoela.* Leipzig: W. Engelmann.

Grüss, J. 1912. *Biologie und Kapillaranalyse der Enzyme.* Berlin: Bornträger.

Guilliermond, A. 1914. "Bemerkungen über die Mitochondrien der vegetativen Zellen und ihre Verwandlung in Plastiden." *Berichte der Deutschen Botanischen Gesellschaft* 32: 282–301.

———. 1921. "La constitution morphologique du cytoplasme dans le cellule végétale." *Revue générale des sciences pures et appliquées* 32: 133–140.

Guilliermond, A., and G. Mangenot. 1922. "Sur la signification de l'appareil réticulaire de Golgi." *Comptes rendus de l'Académie des Sciences de Paris* 174: 692–694.

Haberlandt, G. 1891. "Über den Bau und die Bedeutung der Chlorophyllzellen von *Convoluta roscoffensis.*" Appendix. In L. von Graff, *Die Organisation der Turbellaria acoela.* Leipzig: W. Engelmann.

Hamann, O. 1882. "Zur Entstehung und Entwickelung der grünen Zellen bei *Hydra.*" *Zeitschrift für wissenschaftliche Zoologie* 37: 457–464.

Hariot, P. A. 1892. "Sur une algue qui vit dans les racines des Cycadées." *Comptes rendus de l'Académie des Sciences de Paris* 115: 325.

Harrison, R. G. 1903. "Experimentelle Untersuchungen über die Entwicklung der Seitenlinie bei den Amphibien." *Archiv für mikroskopische Anatomie und Entwicklungsgeschichte* 63: 35–149.

———. 1907. "Observations on the living developing nerve fiber." *Proceedings of the Society for Experimental Biology and Medicine* 4: 140–143.

Harz, C. D. 1885. *Landwirthschaftliche Samenkunde.* 2 vols. Berlin: Paul Parey.

Haupt, A. W. 1923. "Cell structure and cell division in the Cyanophyceae." *Botanical Gazette* 75: 170–190.

Hayem, G. 1877. "Sur l'évolution des globules rouges dans le sang des vertèbres ovipares." *Comptes rendus de l'Académie des Sciences* 85: 907–909.

Heller, H. H. 1921. "Phylogenetic position of the bacteria." *Botanical Gazette* 72: 390–396.

Hellmann, G. 1913. "Über die im Exkretionsorgan der Ascidien der Gattung *Caesira (Molgula)* vorkommenden Spirochäten: *Spirochaeta Caesirae septentrionalis* n. sp. und *Spirochaeta Caesirae retortiformis* n. sp." *Archiv für Protistenkunde* 29: 22–38.

Hertwig, O. 1893–1898. *Die Zelle und die Gewebe.* 2 vols. Jena: G. Fischer.

Hertwig, R. 1902. "Die Protozoen und die Zelltheorie." *Archiv für Protistenkunde* 1: 1–40.

Huxley, T. H. 1858. "On the agamic reproduction and morphology of aphids." *Transactions of the Linnean Society of London* 22: 193–236.

Issajew, W. 1923. "Vererbungsstudien an tierischen Chimären." *Biologisches Zentralblatt* 43: 115–123.

Itzigsohn, E. F. H. 1868. "Kultur der Glaukogonidien von *Peltifera canina.*" *Botanische Zeitung* 1868: 185–196.

Jeffrey, E. C. 1921. *The Anatomy of Woody Plants.* Chicago: University of Chicago Press.

Jönsson, B. 1894. "Studier öfver algparasitism hos *Gunnera* L." *Botaniska Notiser* 1894: 1–20.

Jordan, E. O. 1922. *A Textbook of General Bacteriology.* Philadelphia: W. B. Saunders.

Julin, C. 1912. "Recherches sur le développement embryonnaire de *Pyrosoma giganteum* Les." *Zoologische Jahrbücher,* suppl. 15: 775–863.

Kammerer, P. 1907. "Symbiose zwischen Libellenlarve und Fadenalge." *Archiv für Entwicklungsmechanik der Organismen* 25: 52–81.

———. 1908. "Symbiose zwischen *Oedogonium undulatum* mit Wasserjungferlarven." In *Wiesner-Festschrift.* Im Auftrage des Festkomitees redigiert v. K. Linsbauer, pp. 237–252. Wien: C. Konegen.

Karawajew, W. 1899. "Über Anatomie und Metamorphose des Darmkanals der Larve von *Anobium paniceum*." *Biologisches Centralblatt* 19: 122–130, 161–171, 196–202.

Karpov, V. P. 1922. "Analysis of the cell" [in Russian]. *Yekaterinoslavsky meditsinskii zhurnal* (Yekaterinoslav journal of medicine).

Keller, B. A. 1924. *Obshchaya botanika* (General botany), vol. 1. Voronezh: Kommuna.

Kirchensteins, A. 1922. "Structure intérieure et mode de developpement des bactéries." *Latvijas augstskolas raksti (Acta Universitatis Latviensis)* 3: 1–90, 6 pl.

Knudson, L. 1922. "Non-symbiotic germination of orchid seeds." *Botanical Gazette* 73: 1–25.

Kohl, F. 1903. *Ueber die Organisation und Physiologie der Cyanophyceenzellen und die mitotische Teilung ihres Kernes.* Jena: G. Fischer.

Kolli, A. A. 1894. "The role of microorganisms from a chemical viewpoint" [in Russian]. *Rech' na IX syezde yestestvoispytatelei i vrachei* (Speech at the Ninth Congress of Naturalists and Physicians). Moscow.

Kowalewsky, A. 1901. "Etude biologique de l'*Haementeria costata* Müller." *Mémoires de l'Académie Impériale des Sciences de St.-Pétersbourg,* ser. 8, 11: 1–13.

Kozo-Polyansky, B. M. 1921a. "The theory of symbiogenesis and 'pangenesis, a provisional hypothesis'" [in Russian]. In *Dnevnik I Vserossiiskogo syezda russkikh botanikov v Petrograde v 1921 godu* (Journal of the First All-Russian Congress of Russian Botanists in Petrograd in 1921). Petrograd, p. 101.

———. 1921b. "Symbiogenesis in the evolution of the plant world" [in Russian]. *Vestnik opytnogo dela* [*Bulletin of Experimental Methods*] (Voronezh) 4: 1–24.

———. 1923. *Poslednee slovo anti-Darvinizma* (The Last Word of Anti-Darwinism). Krasnodar: Burevestnik.

Kunstler, J. 1889. "Recherches sur la morphologie des flagelles." *Bulletin scientifique de la France et de la Belgique* 20: 399–515.

Kusano, S. 1911. "*Gastrodia elata* and its symbiotic association with *Armillaria mellea*." *Journal of the College of Agriculture, Japan* 9: 1–73.

Küster, E. 1911. "Über amöboide Formveranderungen der Chromatophoren höherer Pflanzen." *Berichte der deutschen botanischen Gesellschaft* 29: 362–369.

———. 1913. "Zelle und Zellteilung. Botanisch." In *Handwörterbuch der Naturwissenschaften,* vol. 10, pp. 748–807. Jena: G. Fischer.

Lange, F. A. 1881. *Istoriya materializma i kritika ego znacheniya v nastoyashchee vremya* (History of materialism and critique of its current significance), vol. 1. St. Petersburg: Izd-vo L. F. Panteleeva. (Translation of *Geschichte des Materialismus und Kritik seiner Bedeutung in der Gegenwart.*)

Lauterborn, R. 1906. "Zur Kenntnis der sapropelischen Flora." *Allgemeine Botanik* 2: 196–197.

Le Dantec, F. 1911. *Eléments de philosophie biologique.* Paris: Alcan.

Leitgeb, H. 1878. "Die Nostoccolonien im thallus der Anthoceroteen." *Sitzungsberichte der Kaiserlichen Akademie der Wissenschaften in Wien. Mathematisch-naturwissenschaftliche Klasse* 77: 411–418.

Lewis, M. R., and W. H. Lewis. 1911. "The growth of embryonic chick tissues in artificial media, agar and bouillon." *Bulletin of Johns Hopkins Hospital* 22: 126–127.

Lieske, R. 1922. "Bakterien und Strahlenpilze." In K. Linsbauer, ed. *Handbuch der Pflanzenanatomie,* div. 2, pt. 1. Thallophyten, vol. 6. Berlin: Bornträger.

Life, A. C. 1901. "The tuber-like rootlets of *Cycas revoluta.*" *Botanical Gazette* 31: 265–271.

Linsbauer, K. 1917. "Symbiose." In *Illustriertes Handwoerterbuch der Botanik, 2, voellig umgearbeitete Auflage,* ed. C. K. Schneider. Leipzig: W. Engelmann.

Litardière, R. de. 1921. "Recherches sur l'element chromosomique dans la caryocinese somatique des Filicinees." *Cellule* 31: 255–473.

Lumière, A. 1919. *Le myth de symbiotes.* Paris: Masson et cie.

Lundegardh, H. 1922. "Zelle und Cytoplasma." In K. Linsbauer, ed. *Handbuch der Pflanzenanatomie,* div. 1, pt. 1. Cytologie, vol. 1. Berlin: Bornträger.

Lyubimenko, V. N. 1916. "On transformations of plastid pigments in living plant tissue" [in Russian]. *Mémoires de l'Académie Impériale des Sciences,* ser. 8, 33: 1–274.

———. 1918. "On the issue of independence of plastids" [in Russian]. *Zhurnal Russkogo Botanichesgkogo Obshchestva (Journal de la Société Botanique de Russie)* 2: 46–56.

MacDougal, D. T. 1899. "Symbiotic saprophytism." *Annals of Botany* 13: 1–47.

Magnus, W. 1900. "Studien an der endotrophen *Mycorrhiza* von *Neottia Nidus avis.*" *Jahrbücher für wissenschaftliche Botanik* 35: 205–272.

———. 1906. "Über die Formbildung der Hutpilze." *Archiv für Biontologie* 1: 85–161.

Maksimov, A. A. 1914. *Osnovy gistologii. I. Ucheniye o kletke* (Foundations of histology. I. The concept of cell.) Petrograd: K. L. Rikker.

Marchesetti, C. de. 1884. "Sur un nuovo caso di simbiosi." *Atti del Museo Civico di Storia Naturale di Trieste* 7: 239–244.

Martin, G. W. 1922. "The mycoplasm theory." *Botanical Gazette* 74: 337.

Mechnikov, I. I. [Metschnikoff, E.] 1909. *Etyudy optimizma* (Etudes of optimism). 2nd ed. Moscow: Nauchnoe slovo.

Mercier, L. 1906. "Le corps bactéroïdes de la Blatte *(Periplaneta orientalis): Bacillus cuenoti* n. spec." *Comptes rendus de la Société de Biologie* 61: 682.

———. 1911. "Bactéries des invertébrés. I : Les cellules uriques du *Cyclostome* et leur bactérie symbiote." *Bulletin scientifique de la France et de la Belgique* 43.

———. 1913. "Bactéries des Invertébrés. II : La glande a concretion de *Cyclostoma elegans* Drap." *Bulletin scientifique de la France et de la Belgique* 45: 15–26.

[Merezhkovsky, K. S.] Mereschkowsky, C. 1905. "Über die Natur und Ursprung der Chromatophoren im Pflanzenreiche." *Biologisches Centralblatt* 25: 593–604. (English translation: *European Journal of Phycology* 1999, 34: 287–295–Ed.)

———. 1909a. *Teoriya dvukh plazm kak osnova simbiogenezisa, novogo ucheniya o proiskhozhdenii organizmov* (The theory of two plasms as the foundation of symbiogenesis, a new concept of the origin of organisms.) Kazan': Publishing Office of the Imperial University.

———. 1909b. *Konspektivnyi kurs sporovykh rastenii* (A brief course on cryptogamic plants), vols. 1–2. Kazan': Publishing Office of the Imperial University.

[———]. 1910. "Theorie der zwei Plasmaarten als Grundlage der Symbiogenesis, einer neuen Lehre von der Enstehung der Organismen." *Biologisches Centralblatt* 30: 278–367.

Meves, F. 1918. "Die Plasmosomentheorie der Vererbung." *Archiv für mikroskopische Anatomie* 92, II: 41–136.

Meyer, A. 1920. *Morphologische und physiologische Analyse der Zelle der Pflanzen und Thiere.* Jena: G. Fischer.

Meyer, F. J. 1923. "Das trophische Parenchym. A. Assimilationsgewebe." In K. Linsbauer, ed. *Handbuch der Pflanzenanatomie,* Allgemeiner Teil. Histologie., vol. 4, pt. 7A. Berlin: Bornträger.

Miehe, H. 1911. "Die Bakterienknoten an den Blattrandern der *Ardisia crispa* A. DC. Jav. Studien V." *Abhandlungen der mathematisch-physischen Klasse der Königlich Sachsischen Gesellschaft der Wissenschaften* 4: 399–432.

———. 1914. "Weitere Untersuchungen über die Bakteriensymbiose bei *Ardisia crispa.* I. Die Mikroorganismen." *Jahrbücher für wissenschaftliche Botanik* 53: 1–54.

———. 1919. "Weitere Untersuchungen über die Bakteriensymbiose bei *Ardisia crispa.* II. Die Pflanze ohne Bakterien." *Jahrbücher für wissenshaftliche Botanik* 58: 29–65.

———. 1923. "Sind ultramikroskopische Organismen in der Natur verbreitet?" *Biologisches Centralblatt* 43: 1–15.

Mohl, H. von. 1851. *Die vegetabilische Zelle.* Braunschweig.

Molisch, H. 1918. "Das Chlorophyllkorn als Reduktionsorgan." *Sitzungsberichte der Kaiserlichen Akademie der Wissenschaften in Wien. Mathematisch-naturwissenschaftliche Klasse, Abt. I,* 127: 449–472.

Monteverde, N. A., and B. V. Perfiliev. 1914. "On a chlorophyll-like pigment of 'green bacteria' *Pelodictyon*" [in Russian]. *Zhurnal mikrobiologii* (Journal of microbiology) 1: 199–207.

Nadson, G. A., and S. M. Visloukh. 1923. "Structure and biology of the giant bacterium, *Achromatium oxaliferum* Schew" [in Russian]. *Izvestiya Botanicheskogo Sada RSFSR (Bulletin de Principal Jardin Botanique de la Republique Russe)* 22 (suppl. 1): 1–24, 33–37.

Neger, F. W. 1913. *Biologie der Pflanzen auf experimenteller Grundlage (Bionomie).* Stuttgart: F. Enke.

Němec, B. 1901. "Die Bedeutung der fibrillären Strukturen bei den Pflanzen." *Biologisches Centralblatt* 21: 529–538.

———. 1910. *Das Problem der Befruchtungs-Vorgange und andere zytologische Fragen.* Berlin: Bornträger.

Noack, K. L. 1921. "Untersuchungen fiber die individualität der Plastiden bei Phanerogamen." *Zeitschrift für Botanik* 13: 1–35.

Nuttall, G. H. 1923. "Symbiosis in animals and plants." *Report of the British Association for the Advancement of Science,* 197–214.

Oersted, A. S. 1873. *System der Pilze, Lichenen und Algen.* Leipzig: W. Engelmann.

Oes, A. 1913. "Über die Assimilation des freien Stickstoffs durch *Azolla.*" *Zeitschrift für Botanik* 5: 145–163.

Oltmanns, F. 1922–1923. *Morphologie und Biologie der Algen,* vols. 1–2. Jena: G. Fischer.

Oudemans, C. A. J. A. 1872. "Over een byzondere soort van bulzen in der Vlierstam (*Sambucus nigra* L.) tot hiertoe voor een fungus (*Rhizomorpha parallela* Roberge)." *Verslagen en Mededeelingen der Koninklige Akademie van Wetenschappen* 2: 209–229.

Pascher, A. 1914. "Über Symbiosen von Spaltpilzen und Flagellaten mit Blaualgen." *Berichte der deutschen botanischen Gesellschaft* 32: 339–352.

Paulsen, F. 1904. *Vvedenie v filosofiyu* (Introduction to philosophy). Moscow: Izd. Moskovskago Psikhologicheskago Obshchestva. (Translation of *Einleitung in die Philosophie*).

Peklo, J. 1912. "Über symbiontische Bakterien der Aphiden." *Berichte der deutschen botanischen Gesellschaft* 80: 416–419.

———. 1913. "Über die Zusammensetzung der sogenannten Aleuronschicht." *Berichte der deutschen botanischen Gesellschaft* 31: 370–384, table 16.

Penard, E. 1902. *Faune Rhizopodique du Bassin du Léman.* Geneva: Librarie de L'Institut.

Perfiliev, B. V. 1914a. "The chlorophyll-bearing microbe, *Pelodictyon clathrariforme,* of the green bacteria group" [in Russian]. *Zhurnal mikrobiologii* (Journal of microbiology) 1: 197.

———. 1914b. "On the theory of symbiosis of *Chlorochromatium aggregatum* Lauterb. and *Cylindrogloea bacterifera* n.g. n.sp." [in Russian]. *Zhurnal mikrobiologii* (Journal of microbiology) 1: 222–224.

———. 1914c. "On the concept of symbiosis" [in Russian]. *Mikrobiologiya* (Microbiology) 1: 209–224.

Pfeffer, W. 1881. *Pflanzenphysiologie.* Leipzig: W. Engelmann.

Pierantoni, U. 1910. "Ulteriori osservazioni sulla simbiosi ereditaria degli Omotteri." *Zoologischer Anzeiger* 36: 96–111.

———. 1913. "Struttura ed evoluzione dell'organo simbiotico di *Pseudococcus citri* Risso, e ciclo biologico del *Coccidomyces dactylopii* Buchner." *Archiv für Protistenkunde* 31: 300–316.

————. 1914. "La luce degli insetti luminosi e la simbiosi ereditaria." *Rendiconti della Reale Accademia delle Scienze Fisiche e Matematiche di Napoli*, 15–21.

————. 1918. "Gli organi simbiotichi e la luminescenza batterica dei Cephalopodi." *Pubblicazioni della Stazione Zoologica di Napoli* 2: 105–146.

Politis, J. 1921. "Du role du chondriome dans la défense des organismes vegetaux contre l'invasion par les parasites." *Comptes rendus de l'Académie des Sciences de Paris* 173: 421–423.

Portier, P. 1911. *Recherches physiologiques sur les champignons entomophytes*. Thèse présentées à la faculté des sciences de l'Université de Paris pour obtenir le grade de Docteur es Sciences Naturelles. Paris: A. Schultz.

————. 1918. *Les symbiotes*. Paris: Masson et cie.

Pospelov, V. P. 1922. "Infertility in Lepidoptera and an attempt to explain it" [in Russian]. *Izvestiya otdela prikladnoi entomologii Gosudarstvennogo Instituta Opytnoi Entomologii* (Proceedings of the Division of Applied Entomology of the State Institute for Experimental Entomology) 2: 923.

Pratje, A. 1920. "Die Chemie des Zellkernes." *Biologisches Centralblatt* 40: 88–112.

Prenant, A. 1913. "Sur l'origine mitochondriale des granules de pigment." *Comptes rendus de la Société Biologique* 74: 926–929.

Pringsheim, E. G. 1915. "Über das Zusammenleben von Tieren und Algen." *Zeitschrift für Naturwissenschaft* 1915: 26–28.

Randolph, L. F. 1922. "Cytology of chlorophyll types of maize." *Botanical Gazette* 73: 337–375.

Rayner, M. C. 1913. "The ecology of *Calluna vulgaris*. I." *New Phytologist* 12: 59–77.

————. 1915. "Obligate symbiosis in *Calluna vulgaris*." *Annals of Botany* 29: 97–98.

————. 1916. "Recent developments in the study of endotrophic mycorrhiza." *New Phytologist* 15: 161–175.

————. 1921. "The ecology of *Calluna vulgaris*. II. The calcifuge habit." *Journal of Ecology* 9: 60–74.

————. 1923. "Mycorrhiza in the Ericaceae." *Transactions of the British Mycological Society* 8: 61–66.

Regaud, C. 1919. "Mitochondries et symbiotes." *Comptes rendus de la Société Biologique* 82: 309–312.

Reichenow, E. 1910. "*Haemogregarina stepanowi*. Die Entwicklugsgeschichte einer Haemogregarine." *Archiv für Protistenkunde* 20: 251–350.

————. 1922. "Intracelluläre Symbionten bei blutsaugenden Milben und Egeln." *Archiv für Protistenkunde* 45: 95–116.

Reimnitz, J. 1909. *Morphologie und Anatomie von* Gunnera magellanica *Lam*. Ph.D. diss. Kiel.

Reinke, J. 1872. "Parasitische *Anabaena* in Wurzeln der Cycadeen." *Göttingen Nachrichten* 57: 107.

————. 1873. *Morphologische Abhandlungen. II. Untersuchung über die Morphologie der Vegetationsorgane von* Gunnera. Leipzig: W. Engelmann.

————. 1901. *Einleitung in die theoretische Biologie.* Berlin: Schuepp.

Rexhausen, L. 1920. "Über die Bedeutung der ektotrophen Mykorrhiza für die höheren Pflanzen." *Beiträge zur Biologie der Pflanzen* 14: 19–58.

Romieu, M. 1911. "Le spermiogénèse chez l'*Ascaris megalocephala.*" *Archiv für Zellforschung* 6: 254–325.

Rotert, V. A. 1891. *Kurs fiziologii rastenii* (A course of plant physiology). Kazan'.

Roux, W. 1881. *Der Kampf der Theile im Organismus.* Leipzig: W. Engelmann.

Sapĕhin, A. A. 1913. "Untersuchungen über die Individualität der Plastide." *Berichte der Deutschen Botanischen Gesellschaft* 3: 14–16.

Schaudinn, F. 1904. "Generations- und Wirtswechsel bei *Trypanosoma* und Spirochaete." *Arbeiten aus dem Kaiserlichen Gesundheit* 20: 387–439.

Schiller, J. 1907. "Beiträge zur Kenntnis der Entwicklung der Gattung *Ulva.*" *Sitzungsberichte der Kaiserlichen Akademie der Wissenschaften in Wien. Mathematisch-naturwissenschaftliche Klasse* 116: 1691–1716.

Schimper, A. 1885. "Untersuchungen über die Chlorophyllkörper und die ihnen homologen Gebilde." *Jahrbücher für wissenschaftliche Botanik* 16: 1–247.

Schmidt, P. Yu. 1920. *Zagadki zhizni* (The mysteries of life). Petrograd: Gosizdat.

————. 1922. *Sila zhizni* (The force of life). Petrograd: Seyatel'.

Schmitz, F. 1883. "Die Chromatophoren der Algen." *Verhandlungen des naturhistorischen Vereines der Preussischen Rheinlande und Westphalens* 40: 1–180.

Schrader, F. 1921. "The chromosomes of *Pseudococcus nipae.*" *Biological Bulletin* 40: 259–270.

————. 1923. "The origin of the mycetocytes in *Pseudococcus.*" *Biological Bulletin* 45: 279–302.

Schultze, M. 1851. *Beiträge zur Naturgeschichte der Turbellarien.* Greifswald: C. A. Koch.

Schulze, F. E. 1879. "Untersuchungen über den Bau und die Entwicklung der Spongien. Sechste Mittheilung. Die Gattung *Spongelia.*" *Zeitschrift für wissenschaftliche Zoologie* 32: 117–157.

Schürhoff, P. 1908. "Ozellen und Lichtkondensoren bei einigen Peperomien." *Beihhefle zum botanisches Zentralblatt* 23: 14–26.

Senn, G. 1908. *Die Gestalts- und Lageveränderung der Pflanzen-Chromatophoren.* Leipzig: W. Engelmann.

Sharp, L. W. 1914. "Spermatogenesis in *Marsilea.*" *Botanical Gazette* 58: 419–432.

Siegel, J. 1903. "Die geschlechtliche Entwicklung von *Haemoproteus stepanowi* im Rüsselegel *Placobdella catenigera.*" *Archiv für Protistenkunde* 2: 339–342.

Signoret, V. 1868. "Essai monographique sur les Aleurodes." *Annales de la Société Entomologique de France* (4), 8: 369–402.

Sinitsin. 1923. *Etyudy po istorii biologicheskogo determinizma* (Etudes on the history of biological determinism).

Smith, A. L. 1921. *Lichens.* Cambridge: Cambridge University Press.

Smith, E. F. 1921. *An Introduction to Bacterial Diseases of Plants.* Philadelphia: W. B. Saunders.

Spratt, E. R. 1915. "The root nodules of the Cycadaceae." *Annals of Botany* 29: 619–625.

Stahl, E. 1900. "Der Sinn der Mycorhizenbildung." *Jahrbücher für wissenschaftliche Botanik* 34: 539–668.

Stieve, H. 1923. "Neuzeitliche Ansichten über die Bedeutung der Chromosomen, unter besonderer Berücksichtigung der *Drosophila*-Versuche." *Ergebnisse der Anatomie und Entwicklungsgeschichte* 24: 574–577.

Stoward, F. 1911. "A research into the amyloclastic secretory capacities of the embryo and aleurone layer of *Hordeum* with special reference to the question of the vitality and auto-depletion of the endosperm." *Annals of Botany* 25: 799–842, 1147–1204.

Strindberg, M. 1913. "Embryologische Studien an Insekten." *Zeitschrift für wissenschaftliche Zoologie* 106: 1–127.

Sukhov. 1914. "Old and new theories about mycorrhiza" [in Russian]. *Lesnoi zhurnal* (Forest journal).

Šulc, K. 1910. "'Pseudovitellus' und ähnliche Gewebe der Homopteren sind Wohnstätten symbiotischer Saccharomyceten." *Sitzungsberichte der Kaiserlichen Akademie der Wissenschaften in Wien. Mathematisch-naturwissenschaftliche Klasse* 3: 1–39.

Szafer, W. 1911. "Zur Kenntnis der Schwefelflora in der Umgebung von Lemberg." *Bulletin de l'Académie des Sciences de Cracovie, Série B:* 160–167.

Tischler, G. L. 1917. In C. K. Schneider, ed. *Illustriertes Handwoerterbuch der Botanik, 2, voellig umgearbeitete Auflage.* Leipzig: Engelmann.

———. 1922. "Allgemeine Pflanzenkaryologie." In K. Linsbauer, ed. *Handbuch der Pflanzenanatomie,* div. 1, pt. 1, vol. 2. Berlin: Bornträger.

Tobler, F. 1911. "Zur Biologie von Flechten und Flechtenpilzen I." *Jahrbücher für wissenschaftliche Botanik* 49: 389–409.

Trojan, E. 1919. "Bakteroiden, Mitochondrien und Chromidien." *Archiv für Mikroskopische Anatomie,* sec. 1, 93: 333–374.

Tubeuf, K. F., von. 1897. *Diseases of Plants Induced by Cryptogamic Parasites.* London: Longmans, Green.

Unger, F. 1846. *Grundzüge der Anatomie und Physiologie der Pflanzen.* Wien: Carl Gerold.

Vainio, E. A. 1921. "Mycosymbiose. Symbiose de deux Champignons." *Annales Botanici Societatis Zoologicae Botanicae Fennicae Vanamo* 1: 56–60.

Vavilov, N. I. 1916. "A survey of the concept of transplantation" [in Russian]. *Sad i ogorod* (Garden and vegetable garden) 32: 2–3.

Vogl, A. 1898. "Mehl und die anderen Mehlproduckte der Cerealien und Leguminosen." *Zeitschrift für Nahrungsmittel-Untersuchung, Hygiene und Waarenkunde* 12: 25–29.

Wallin, I. E. 1922a. "On the nature of mitochondria. I. Observations on mitochondria staining nethods applied to bacteria. II. Reactions of bacteria to chemical treatment." *American Journal of Anatomy* 30: 203–229.

————. 1922b. "On the nature of mitochondria. III. The demonstration of mitochondria by bacterial methods. IV. A comparative study of the morphogenesis of root-nodule bacteria and chloroplasts." *American Journal of Anatomy* 30: 451–471.

————. 1923. "On the nature of mitochondria. V. A critical analysis of Portier, *Les symbiotes.*" *Anatomical Record* 25: 1–7.

Warming, E. 1901. *Oikologicheskaya geografiya rastenii* (Oecological plant geography). Moscow: Tipografiya I. A. Balandina (Translation of *Plantesamfund; Grundtraek af den økologiske Plantegeografi*).

Webber, H. J. 1897. "Notes on the fecundation of *Zamia* and the pollen tube apparatus of *Ginkgo.*" *Botanical Gazette* 24: 225–235.

————. 1901. "Spermatogenesis and fecundation of *Zamia.*" *USDA Bureau of Plant Industry Bulletin* 2: 1–100.

Weismann, A. 1918. *Lektsii po evolyutsionnoi teorii* (Lectures on the evolutionary theory). Petrograd: A. F. Devrien (Translation of *Vorträge über Deszendenztheorie*).

Weiss, A. 1864–1866. "Untersuchungen über die Entwickelungsgeschichte des Farbstoffes in Pflanzenzellen." *Sitzungsberichte der Kaiserlichen Akademie der Wissenschaften in Wien. Mathematisch-naturwissenschaftliche Klasse, Abt. I,* 1864, 50(1): 6–35; 1866, 54(1): 157–217.

Weiss, E. 1884. "Beitrag zur Culm-Flora von Thüringen." *Jahrbuch Preussische Geologische Landesanstalt* 1883: 81–100.

Welsford, E. J. 1915. "Nuclear migrations in *Phragmidium violaceum.*" *Annals of Botany* 29: 293–298.

West, G. S. 1916. "Algae (Myxophycecae, Peridineae, Bacillariae, Chlorophyceae)." In *Cambridge Botanical Handbooks,* vol. 1, ed. A. C. Seward and A. G. Tansley. Cambridge: Cambridge University Press.

Wettstein, F., von. 1921. "Das Vorkommen von Chitin und seine Verwertung als systematisch-phylogenetisches Merkmal in Pflanzenreich." *Sitzungsberichte der Kaiserlichen Akademie der Wissenschaften in Wien. Mathematisch-naturwissenschaftliche Klasse, Abt. I,* 130: 3–20.

Wettstein, R. 1923. *Handbuch der systematischen Botanik,* 3rd ed., A, vol. 1, Leipzig and Vienna: Franz Deuticke.

Whetzel, H. H. 1918. *An Outline of the History of Phytopathology.* Philadelphia: W. B. Saunders.

Whewell, W. 1840. *The Philosophy of the Inductive Sciences, Founded upon Their History.* London: Parker.

Wigand, A. 1887. "Die Krystallnatur des Plastiden." *Botanische Hefte, Marburg,* 2.

————. 1888. "Das Protoplasma als Fermentorganismus." *Botanische Hefte, Marburg,* 3.

Winckler, H. 1907. "Über Pfropfbastarde und pflanzliche Chimären." *Berichte der Deutschen Botanischen Gesellschaft* 25: 567–576.

Winogradsky, S. 1895. *Microbiologie du sol.* Paris: Mason et cie.

Witlaczil, E. 1882. "Zur Anatomie der Aphiden." *Arbeiten aus dem Zoologischen Institut der Universität Wien* 4: 397–441.

Wülker, G. 1913. "Über das Auftreten rudimentären akzessorischer Nidamentaldrüsen bei männlichen Cephalopoden." *Zoologica* 67 (Bd 26): 201–210.

Kozo-Polyansky's Taxa

This list includes all the Latin names of organisms mentioned by Kozo-Polyansky (if the name was given only in Russian, the Latin name is listed in brackets), with their modern synonymy (if applicable) and abbreviated modern placement/classification. "Higher" (most inclusive) taxa after Margulis and Chapman (2010).

Name	Classification
Abutilon	Plantae: Anthophyta: Malvaceae
Acetobacter acetixylinus (as *Bacterium xylinum*)	Prokaryotae: Proteobacteria: Alphaproteobacteria: Rhodospirillales: Acetobacteraceae
Aeschna	Animalia: Mandibulata: Insecta: Odonata: Aeschnidae
Aleurodidae	Animalia: Mandibulata: Insecta: Homoptera
Amylomyces rouxii	Fungi: Zygomycota
Anabaena	Prokaryotae: Cyanobacteria: Nostocales
Anabaena azollae	see *Anabaena*
Anobiinae	Animalia: Mandibulata: Insecta: Coleoptera: Anobiidae
Anthoceros	Plantae: Anthocerophyta: Anthocerotaceae
Anthoceros laevis	see *Anthoceros*
Aphidae	Animalia: Mandibulata: Insecta: Homoptera
Aplectrum	Plantae: Anthophyta: Orchidaceae
Ardisia crispa	Plantae: Anthophyta: Myrsinaceae
Armillaria mellea	Fungi: Basidiomycota: Agaricales: Marasmiaceae
Asteraceae (as Compositae)	Plantae: Anthophyta
Azolla	Plantae: Filicinophyta: Salviniales: Azollaceae

Azotobacter	Prokaryotae: Proteobacteria: Gammaproteobacteria: Pseudomonadales
Bacillus cuenoti	see *Blattabacterium cuenoti*
Bacillus mycoides	Prokaryotae: Endospora: Bacillales: Bacillaceae
Bacillus radicicola, see *Rhizobium radicicola*	Prokaryotae: Proteobacteria: Rhizobiales
Bacterium polychromum	Prokaryotae: ?Proteobacteria: ?Chromatiaceae
Bacterium xylinum	see *Acetobacter acetixylinus*
Batrachospermum	Protoctista: Rhodophyta: Batrachospermales
Blasia	Plantae: Hepatophyta: Hepaticae
Blattabacterium cuenoti (as *Bacillus cuenoti*)	Prokaryotae: Enterobacteriales: Enterobacteriaceae
Cactaceae	Plantae: Anthophyta
Calloriopsis	Fungi: Ascomycota: Helotiaceae
Calluna vulgaris	Plantae: Anthophyta: Ericaceae
Camponotus	Animalia: Mandibulata: Insecta: Hymenoptera: Formicidae
Carteria	Protoctista: Chlorophyta: Volvocales: Carteriaceae
Cephaleuros virescens	Protoctista: Chlorophyta: Trentepohliales: Trentepohliaceae
Ceratodictyon spongiosum (as *Marchesettia spongioides*)	Protoctista: Rhodophyta: Rhodymeniales: Gracilariaceae
Chaetophora	Protoctista: Chlorophyta: Chaetophorales: Chaetophoraceae
Chamaecytisus hirsutus (as *Cytisus hirsutus*)	Plantae: Anthophyta: Fabaceae
Chamaecytisus purpureus (as *Cytisus purpureus*)	Plantae: Anthophyta: Fabaceae
Chlorella vulgaris	Protoctista: Chlorophyta: Chlorococcales: Chlorellaceae
Chlorobium (as *Pelodictyon*)	Prokaryotae: Chlorobium: Chlorobiales
Chlorochromatium	a consortium of Prokaryotae (Proteobacteria and Chlorobium)
Chlorochromatium aggregatum (also as *Chloronium mirabile*)	see *Chlorochromatium*
Chloronium mirabile	see *Chlorochromatium aggregatum*
Chroostipes	Prokaryotae: Cyanobacteria: Chroococcales
Cicada orni	Animalia: Mandibulata: Insecta: Homoptera: Cicadidae
Cicadidae	Animalia: Mandibulata: Insecta: Homoptera

Clostridium	Prokaryotae: Firmicutes: Clostridiales: Clostridiaceae
Clostridium pasterianum	see *Clostridium*
[*Coccus cacti*]	Animalia: Mandibulata: Insecta: Hemiptera: Monophlebidae
Conferva	see n. 27 to ch. 3
Convoluta roscoffensis	see *Symsagittifera roscoffensis*
Coprinus sterquilinus	Fungi: Basidiomycota: Agaricales: Coprinaceae
Crataegomespilus	Plantae: Anthophyta: Rosaceae
Crataegomespilus asnieresii	see *Crataegomespilus*
Crataegomespilus dardarii	Plantae: Anthophyta: Rosaceae
Crataegus monogyna	Plantae: Anthophyta: Rosaceae
Cyanodictyon endophyticum	Prokaryotae: Cyanobacteria: Synechococcales: Synechococcaceae
Cyanomonas	Protoctista: Cryptomonada: Cryptomonadaceae
Cyanophyceae	= Cyanobacteria (Prokaryotae)
Cyanotheca longipes	Fungi (incertae sedis)
Cyclostoma elegans	see *Pomatias elegans*
Cylindrogloea bacterifera	a consortium of Prokaryotae (Proteobacteria and Chlorobia)
Cytisus adamii	see *Laburnocytisus adamii*
Cytisus hirsutus	see *Chamaecytisus hirsutus*
Cytisus laburnum	see *Laburnum anagyroides*
Cytisus purpureus	see *Chamaecytisus purpureus*
Cytorreictes	see n.34 to chap. 3
Dactylococcopsis	Prokaryotae: Cyanobacteria: Chroococcales
Dalyellia viridis (as *Vortex viridis*)	Animalia: Platyhelminthes: Turbellaria: Rhabdocoela: Dalyellidae
[*Datura*]	Plantae: Anthophyta: Solanaceae
Decapodiformes	Animalia: Mollusca: Cephalopoda
Dendroceros	Plantae: Anthocerophyta: Dendrocerotaceae
Didymoplexis	Plantae: Anthophyta: Orchidaceae
Diplothrix	see *Calloriopsis*
Echinocactus	Plantae: Anthophyta: Cactaceae
Entomophoraceae	Fungi: Zygomycota: Entomophorales
Ericaceae	Plantae: Anthophyta
Funaria	Plantae: Bryophyta
Gastrodia	Plantae: Anthophyta: Orchidaceae
Gastrodia elata	see *Gastrodia*
Gloeocapsa	Prokaryotae: Cyanobacteria: Chroococcales: Chroococcaceae
Gonidiomyces	Fungi: Ascomycota (incertae sedis)

Gunnera	Plantae: Anthophyta: Gunneraceae
Halichondria	Animalia: Porifera: Demospongiae:
Halichondriidae	
Haliclona fibulata	Animalia: Porifera: Demospongiae: Chalinidae
(as *Reniera fibulata*)	
Hippuris	Plantae: Anthophyta: Hippuridaceae
Hordeum coeleste	Plantae: Anthophyta: Poaceae
Hydra vulgaris	Animalia: Coelenterata: Hydrozoa: Hydridae
Isarieae	Fungi: Ascomycota: Clavicipitaceae
Laburnocytisus adamii	Plantae: Anthophyta: Fabaceae
(as *Cytisus adamii*)	(*Laburnocytisus adamii* = a graft chimera,
	Chamaecytisus purpureus + *Laburnum anagyroides*)
Laburnum anagyroides	Plantae: Anthophyta: Fabaceae
(as *Laburnum vulgare*	
and *Cytisus laburnum*)	
Laburnum vulgare, see	Plantae: Anthophyta: Fabaceae
Laburnum anagyroides	
Lampyridae	Animalia: Mandibulata: Insecta: Coleoptera
Lampyris noctiluca	Animalia: Mandibulata: Insecta: Coleoptera:
	Lampyridae
Lepidoptera	Animalia: Mandibulata: Insecta
Liponyssus saurarum	see *Oudemansiella saurarum*
Loligo	Animalia: Mollusca: Cephalopoda: Loliginidae
Loligo forbesii	see *Loligo*
Lolium	Plantae: Anthophyta: Poaceae
Lolium temulentum	see *Lolium*
Marchesettia spongioides	see *Ceratodictyon spongiosum*
[*Loxostege sticticalis*]	Animalia: Mandibulata: Insecta: Lepidoptera:
	Pyralidae
Merismopedia	Prokaryotae: Cyanobacteria: Chroococcales:
	Chroococcaceae
[*Mespilus*]	Plantae: Anthophyta: Rosaceae
Modiola	Plantae: Anthophyta: Malvaceae
Molgulidae	Animalia: Urochordata
Myrsinaceae	Plantae: Anthophyta
Neottia nidus-avis	Plantae: Anthophyta: Orchidaceae
Nephromyces	Protoctista: Apicomplexa
[*Nosema*]	Fungi: Microspora
Nostoc	Prokaryotae: Cyanobacteria: Nostocales
Nostoc gunnerae	see *Nostoc*
Nostoc punctiforme	see *Nostoc*
Notothylas	Plantae: Anthocerophyta: Notothyladaceae

Oedogonium undulatum	Protoctista: Chlorophyta: Oedogoniales: Oedogoniaceae
Oenothera gigas	Plantae: Anthophyta: Onagraceae
Oikomonas	Protoctista: Chrysomonada: Chromulinales: Oikomonadidae
Oikomonas syncyanotica	Protoctista: Chrysomonada: Chromulinales: Oikomonadidae (a consortium with *Chroostipes linearis*)
Orcheomyces	Fungi: Basidiomycota: Ceratobasidiales
Oscillaria	Prokaryotae: Cyanobacteria: Oscillatoriales
Oudemansiella saurarum (as *Liponyssus saurarum*)	Animalia: Chelicerata: Arachnida: Acari: Macronyssidae
Pavetta angustifolia	Plantae: Anthophyta: Rubiaceae
Pelargonium	Plantae: Anthophyta: Geraniaceae
Pelmatohydra oligactis	Animalia: Coelenterata: Hydrozoa: Hydridae
Pelmatohydra oligactoides	Animalia: Coelenterata: Hydrozoa: Hydridae
Pelochromatium roseum	a consortium of Prokaryotae (Proteobacteria and Chlorobium)
Pelodictyon, see *Chlorobium*	Prokaryotae: Chlorobia: Chlorobiales
Pelomyxa	Protoctista: Arcaeaprotista: Pelobiontae: Pelomyxidae
Peperomia arifolia	Plantae: Anthophyta: Piperaceae
Peperomia metallica	Plantae: Anthophyta: Piperaceae
Peperomia sandersii	Plantae: Anthophyta: Piperaceae
Pereskia	Plantae: Anthophyta: Cactaceae
Peridinea	Protoctista
Phoma	Fungi: anamorphic Ascomycota
Phycomyces nitens	Fungi: Zygomycota: Mucorales
Phyllophoma [invalid name]	Fungi: Ascomycota
Placobdella catenigera	Animalia: Annelida: Hirudinea
Pleospora	Fungi: Ascomycota: Pleosporales
Pomatias elegans (as *Cyclostoma elegans*)	Animalia: Mollusca: Gastropoda: Pomatiasidae
[*Populus nigra*]	Plantae: Anthophyta: Salicaceae
Populus trichocarpa	Plantae: Anthophyta: Salicaceae
Pseudococcus	Animalia: Mandibulata: Insecta: Hemiptera: Pseudococcidae
Pyrosoma	Animalia: Urochordata: Thaliacea: Pyrosomidae
Radiolaria	Protoctista: Actinopoda
Rana palustris	Animalia: Craniata: Amphibia: Anura: Ranidae
Rana sylvestris	Animalia: Craniata: Amphibia: Anura: Ranidae
Reniera fibulata	see *Haliclona fibulata*

Rhizobium	Prokaryotae: Proteobacteria: Rhizobiales
Rhizobium radicicola (as *Bacillus radicicola*)	see *Rhizobium*
Rhizoctonia	Fungi: Basidiomycota: Corticiales
Rhizomorpha	Fungi: incertae sedis
Rondeletiola minor	Animalia: Mollusca: Cephalopoda: Sepiolidae
Rosaceae	Plantae: Anthophyta
Rubiaceae	Plantae: Anthophyta
Saccharomycetes	Fungi: Ascomycota
Salvinia natans	Plantae: Filicinophyta: Salviniales
[*Sambucus nigra*]	Plantae: Anthophyta: Adoxaceae
Schizophyta	= Prokaryotae
Scytonema	Prokaryotae: Cyanobacteria: Nostocales
Sepia	Animalia: Mollusca: Cephalopoda: Sepiidae
Sepia elegans	see *Sepia*
Sepiola	Animalia: Mollusca: Cephalopoda: Sepiolidae
Sepiola elegans	see *Sepiola*
Sepiolidae	Animalia: Mollusca: Cephalopoda
Solanaceae	Plantae: Anthophyta
Solanum	Plantae: Anthophyta: Solanaceae
Solanum gaertnerianum	see *Solanum*
Solanum koelreuterianum	see *Solanum*
Solanum tubingense	see *Solanum*
Sphaerocystis schroeteri	Protoctista: Chlorophyta: Tetrasporales: Tetrasporaceae
Spirogyra	Protoctista: Chlorophyta: Zygnematales: Zygnemataceae
Spongelia	Animalia: Porifera: Demospongiae: Dysideidae
Spongia otochelica [correct: *otahetica*]	Animalia: Porifera: Demospongiae: Spongidae
Spongocladia	Protoctista: Chlorophyta: Cladophorales: Siphonocladaceae
Struvea	Protoctista: Chlorophyta: Cladophorales: Siphonocladaceae
Symsagittifera roscoffensis (as *Convoluta roscoffensis*)	Animalia: Platyhelminthes: Turbellaria: Acoela: Sagittiferidae
Thamnoclonium flabelliforme	Protoctista: Rhodophyta: Florideophyceae: Halymeniales: Halymeniaceae
Ustilaginales	Fungi: Basidiomycota
Ustilago tritici	Fungi: Basidiomycota: Ustilaginales: Ustilaginaceae
Vortex viridis	see *Dalyellia viridis*

Editors' Commentary

Preface

1. Biology too also demonstrates "disintegration of elements"—the ancient atomists' intuition unexpectedly applies to "individual cells" (eukaryotic "elements"). Insight of both Kozo-Polyansky and ancient Greek scholars is bolstered by chemical, ecological, physiological, and other studies that show all eukaryotes to be composed of smaller prokaryotic "atoms."
2. *William Whewell* (1794–1866), English philosopher, polymath, and founder of modern philosophy of science.
3. *Nikolay Ivanovich Vavilov* (1887–1943), famous Russian/Soviet plant geneticist, killed in Stalin's prison. Kozo-Polyansky's own teacher, Professor Nikolay Maksimovich Tulaikov (1875–1938), also perished, along with many other scientists. *Mikhail Ilyich Golenkin* (1864–1941), Russian/Soviet botanist, Moscow University professor and another teacher of Kozo-Polyansky as well as many famous biologists (N. Timofeev-Ressovsky, E. Wulf, etc). *Lev Ivanovich Kursanov* (1877–1954), Russian/Soviet botanist, Moscow University professor, author of textbooks on lower plants. *Vladimir Nikolaevich Lyubimenko* (1873–1937), Russian/Soviet botanist, studied photosynthesis. *Donald Reddick* (1883–1955), plant pathologist, Cornell University (Ithaca, New York). *George W. Martin* (1886–1971), botanist, mycologist, Rutgers University (New Brunswick, New Jersey; Kozo-Polyansky lists "Brunswick"), then Iowa State University. *Jacob R. Schramm* (1885–1976), botanist, St. Louis Botanical Garden and Washington University, St. Louis, Missouri, then Cornell University and the University of Pennsylvania; Kozo-Polyansky lists, incorrectly, "Washington." Judging from the acknowledgments and the ample literature list, Kozo-Polyansky had access to the latest European and American journals of 1921–1923. This is not a trivial circumstance, since until NEP

("new economic politics") was declared by the Bolshevik dictatorship in 1921, contacts of Russian scientists with the West were severely curtailed.

I. Noncellular Organisms (Cytodes) and Bioblasts [Prokaryotes]

1. "Cytode" and "bioblast" are currently obsolete, vaguely synonymous terms for elementary, non-nucleated units of life. *Kozo-Polyansky uses these terms for a non-nucleated* (i.e., prokaryotic) organism (cytode) and its cell (bioblast), as opposed to nucleated (eukaryotic) organisms and cells. By using this terminology, outdated already in his time, *Kozo-Polyansky emphasized the symbiogenetic origin of nucleated organisms and their cells.* The word "cytode" was introduced in 1866 by Ernst Haeckel (1834–1919), and the word "bioblast" in 1870 by Lionel Beale (Sapp 2003: 88). The term "bioblast" was also used for mitochondria by Altmann (1890).

2. Babes bodies, or Babes-Ernst bodies (corpuscles, granules), or metachromatic granules are polyphosphate inclusions in the cytoplasm of gram-positive bacteria such as diphtheria. Kozo-Polyansky calls them in Russian "tel'tsa Baba" ["Bab bodies"], a French reading of the surname Babes. Victor Babeş (1854–1926) was a famous Romanian biologist, one of the early bacteriologists; his last name is correctly pronounced "Babesh."

3. *Pelodictyon* Lauterborn 1913 until recently was a valid genus of green bacteria (Chlorobiales: Chlorobiaceae); however, Imhoff (2003) proposed to transfer its type species of the genus *Pelodictyon* to the genus *Chlorobium.*

4. For current reviews of cyanobacterial symbioses in general, see Adams (2000), Adams et al. (2006), Rai, Bergman, and Rasmussen (2002); on symbioses of cyanobacteria with nitrogen-fixing bacteria, see also Elmerich and Newton (2003).

5. Anabaenin is "cyanophycean starch," a copolymer of aspartic acid and arginine.

6. *Cyanotheca longipes* was excluded from cyanobacteria by Claus (1961), who found a nucleus in this species; therefore, it is a protoctist.

7. *Cyanodictyon* Pascher 1914 (Synechococcales, Synechococcaceae) is a currently valid genus of cyanobacteria (Komárek and Hauer 2004). *C. endophyticum* Pascher 1914 is endogloeic in the mucilage of *Anabaena.*

8. The phototrophic consortium *Chlorochromatium aggregatum* is currently treated as the most highly developed interspecific association of bacteria; it consists of green sulfur bacteria surrounding a central, motile, chemotrophic bacterium (Abella et al. 1998; Wanner, Vogl, and Overmann 2008; Vogl et al. 2006, 2008). For other reviews of phototrophic green bacterial consortia such as *Chlorochromatium* see Kanzler et al. (2005), Overmann (2006), Overmann and Schubert (2002).

9. *Oikomonas syncyanotica,* a chrysophycean protoctist, appears to gain nutrition from its union with symbiotic cyanobacteria ("cyanellae"); its motility occurred in the absence of free oxygen but was light-dependent (Sleigh 1991: 283).

10. On remarkable symbiosis of *Pelomyxa,* see Buchner (1965: 92–94); Margulis (1993). *Pelomyxa* is a unique, giant (500–800 μm) amoeba, lacking mitochondria, endoplasmic reticulum, and Golgi; it contains three types of symbiotic bacteria, two of which are methanogenic archaebacteria.

II. The Cell and Its Organelles

1. Symbioses described in this section refer not only to animals but also to unicellular Protoctista (Protista, "Protozoa"), which Kozo-Polyansky classified traditionally as "lower animals."

2. Zoochlorellae are usually related to chlorophytes such as *Chlorella;* zooxanthellae are dinomastigotes such as *Symbiodinium microadriaticum.* For a review of zooxanthellae symbiotic with octocorals, see, e.g., Van Oppen et al. (2005). For a novel remarkable secondary symbiosis *(Hatena)* between a nonphotosynthetic protoctist and a dinomastigote, see Okamoto and Inouye (2006).

3. On the famous symbiosis of the worm *Symsagittifera* (formerly *Convoluta*) *roscoffensis* (Platyhelminthes: Turbellaria: Acoela), see Buchner (1965: 12); Margulis (1993: 186, 193, 194, table 7–5; 203, 221, table 8–1). The symbiont is green alga *Tetraselmis* (formerly *Platymonas*) *convolutae.* For a review of photosynthetic symbioses in animals, see Venn, Loram, and Douglas (2008). See the back cover on the remarkable mollusc *Elysia* symbiosis (Mujer et al. 1996; Rumpho et al. 2008).

4. Chloroplasts ("chlorophyll organelles"): symbiotic origin of eukaryotic chloroplasts from photosynthetic cyanobacteria is amply proved today (Margulis 1993). For a review, see Reyes-Prieto, Weber, and Bhattacharya (2007).

5. Centrosome/centriole: see note 15 below on blepharoplasts.

6. Mature chloroplast division in *Anthoceros* is mentioned in Margulis (1993: 328); see also Izumi et al. (1993).

7. On nucleus origin, see Margulis, Dolan, and Guerrero (2000); Margulis et al. (2006); and Dolan et al. (2002).

8. On bacterial endosymbioses in mealybugs *Pseudococcus* (Insecta: Homoptera: Coccoidea: Pseudococcidae), see Buchner (165: 254–263). Currently, these remarkable symbioses are under close study: primary symbionts belonging to Betaproteobacteria can be in their turn inhabited by secondary symbionts belonging to Gammaproteobacteria (Thao et al. 2002).

9. On chromidia, see Powell (1979) and Margulis, Enzien, and McKhann (1990). The nature of these enigmatic structures is under discussion; encysting amoebae mastigotes (e.g., *Paratetramitus jugosus*) can reproduce by chromidia.

10. The symbiotic origin of mitochondria from aerobic Alphaproteobacteria is amply proved today (Margulis 1993); see also Andersson et al. (2003) and Searcy (2003).

11. Independent growth of mitochondria in culture, experimentally pursued in particular by Portier (1918) and Wallin (1922, 1927), has not been achieved.

12. The fact that Portier (1918) and Wallin (1927) both claimed that they had grown mitochondria outside cells may have been the major reason that their work was vociferously rejected by research biologists, their reputations compromised and careers curtailed. Persistent demonstration by many biologists from before the work of Kozo-Polyansky to the present supports the current well-founded conclusion. Neither mitochondria nor chloroplasts have ever been cultivated beyond the confines of an intact live cell by that cell. In spite of the many claims to have grown organelles (some even misled master scientists including Lewis Thomas, 1978), Kozo-Polyansky is not correct here. No organelles has ever been supplied with the sufficiency of genes, proteins, metabolites, or vitamins required for growth and reproduction. Mitochondria and chloroplasts have such small, limited genomes relative to the free-living bacteria from which they evolved that to supply them with their needs, so far, has been a feat beyond the skill of any modern scientist. As Kozo-Polyansky rightly recognized, in principle, the growth of an isolated organelle would definitely prove its symbiotic provenance.

13. "Ergastoplasm" is an old term for endoplasmic reticulum, with ribosomes staining. There is little evidence of its symbiotic origin.

14. Physodes are membrane-bound organelles in brown algae, which contain phenolic compounds (phlorotannins), important for algal metabolism and development (Schoenwaelder and Clayton 1999; Schoenwaelder 2002).

15. The blepharoplast (today called kinetosome)-centriole complex has a central role in a current model of eukaryotic cell and nucleus origin (karyomastigont model, spirochaete-archaebacterial symbiogenesis); see Margulis and McMenamin (1990); Margulis (1993); Margulis and Chapman (1998); Sapp (1998); Chapman, Dolan, and Margulis (2000); Margulis, Dolan, and Guerrero (2000); and Margulis et al. (2006). Kozo-Polyansky and Wallin are the only early "symbiogeneticists" who both nonchalantly suggested the symbiogenetic origin of eukaryotic motility organelles.

16. Elaioplast is a modified plastid specialized for lipid storage.

17. There is no modern evidence whatsoever of a symbiotic origin of the aleurone. Kozo-Polyansky seems to rely completely on the data of J. Peklo (1913); on Peklo, see also commentary to p. 49. At the same time, "endophyte" zygomycetous fungi in many grasses are well known. See fig. III-6.

18. Endophyte fungi *Epichloe* (asexual forms = *Neotyphodium* spp.) (Clavicipitaceae), toxic to animals, are symbiotic with many grasses (Poaceae), including

Lolium; for reviews, see Clay and Schardl (2002) and Schardl, Leuchtmann, and Spiering (2004). See also p. 50.

III. Multicellular Organisms

1. For reviews of lichen symbioses, see Honegger (1991), Ahmadjian (1993) and Brodo, Sharnoff, and Sharnoff (2001). The research of early Russian lichenologists (Elenkin, Genkel, and others) is described in detail in Khakhina (1992). While the majority of lichens are formed by bipartite fungus–green algae symbiosis, or cyanobacterial photobionts, some (tripartite or quadripartite) also include actinobacteria, other cyanobacteria, etc. (Adams et al. 2005; Brodo, Sharnoff, and Sharnoff 2001).

2. On chimeras in plants, see, e.g., Marcotrigiano (1997); Chakrabarty, Mandal, and Datta (2000).

3. *Pelmatohydra oligactis:* the specific name in the original Kozo-Polyansky text is misspelled *"olidactis,"* and the chimera described further as "olidactoid."

4. On the modern hypothesis of transplantation as symbiosis, see Chigira (1997).

5. *Thamnoclonium flabelliforme:* the generic name in the original text is misspelled as *"Thamnocladium."* For a modern review of sponge-algal symbiosis, see Trautman and Hinde (2002).

6. Numerous symbioses of cyanobacteria with pteridophytes and bryophytes are now well studied. Well-studied cases include cyanobionts of rice paddy fern *Azolla* and hornwort *Anthoceros* described by Kozo-Polyansky. For current reviews of cyanobacterial-plant symbioses, see Adams et al. (2005); Rai, Bergman, and Rasmussen (2002); and Rai, Söderbäck, and Bergman (2000); for a review of cyanobacterial-bryophyte symbioses, see Adams and Duggan (2008). *Azolla* symbiont apparently does not belong to *Anabaena* or *Nostoc* (Baker, Entsch, and McKay 2003). See fig. III-2.

7. The remarkable symbiosis of cyanobacterium *Nostoc punctiforme* and angiosperm plant *Gunnera* is now investigated in detail; for reviews see Adams et al. (2005), Benson and Margulis (2002), Bergman (2002), Bergman and Osborne (2002), Johansson and Bergman (1994), Nilsson, Bergman, and Rasmussen (2000); and Rai, Söderbäck, and Bergman (2000). The *Gunnera* symbiosis is maintained (in part at least) by N_2 fixation.

8. On eubacterial endosymbionts of Rubiaceae, see Van Oevelen et al. (2002). *Ardisia* symbionts (for their biology, see, e.g., Nakahashi, Frole, and Sack 2005) are still undescribed.

9. In gymnosperm cycads (Cycadaceae) all 150 species in nine genera form symbiogenetic coralloid roots with cyanobacteria; see Rai, Söderbäck, and Bergman (2000). The *Nostoc* symbiont produces β-methylamino-L-alanine (BMA), a toxin found in cycad seeds (Cox et al. 2005).

10. Diverse and ubiquitous symbioses of terrestrial plants with mycorrhizal fungi are well documented today. For reviews, see, e.g., Strack et al. (2003) and Finlay (2008). Apparently, the fungi enhance the ratio of phosphorus and nitrogen transfer to plant roots (Eissenstat 1990; Karandashov and Bucher 2005; Read 2005; Leigh, Hodge, and Fitter 2009). Prokaryotes such as *Frankia* that have hyphae-like extensions provide nitrogen, but no eukaryotes (fungi) fix atmospheric N_2; this is an entirely bacterial process.

11. For a review of orchid-fungus symbioses, see Dearnaley (2007). Most orchid mycorrhizal fungi are currently classified in the "form-genus" *Rhizoctonia* (Basidiomycota: Corticiales).

12. See n. 18 to chapter II.

13. Freshwater algal growth (epizoic symbiosis), common on larvae (naiads) of dragonflies (Odonata) and other aquatic insects (mosquito larvae, etc.), are considered mutualistic (Jolivet 1998: 144).

14. *Camponotus* ants (Insecta: Hymenoptera: Formicidae) famously form symbioorgans in their gut (Buchner 1965: 531–542) with bacterial symbiont *Blochmannia* (Wernegreen et al. 2002; Degnan et al. 2004; de Souza et al. 2009). For a review of insect mycetomes see Douglas (1989); for a general review of bacterial endosymbionts in animals, see Moran and Baumann (2000).

15. Anobiine (Insecta: Coleoptera: Anobiidae) beetles (e.g., *Lasioderma* and *Stegobium;* Buchner 1965: 118–128) have fungal yeastlike endosymbionts (YLS) *Symbiotaphrina* (Noda and Kodama 1996). YLS could originate from within the filamentous ascomycetes (Euascomycetes).

16. See Buchner (1965: 254–262) for early work on "pseudovitellus" in aphids (Insecta: Homoptera: Aphidoidea). "What a wealth of guesswork did this pseudovitellus inspire!" (Buchner 1965: 25). This symbioorgan contains a gamma-proteobacterial endosymbiont, *Buchnera*, remarkable by its reduced genome (Baumann et al. 1998; Douglas 1998, 2003; Shigenobu et al. 2000; Gil et al. 2002; Latorre and Moya 2006; Moran and Degnan 2006). Andersson (2000) even suggested that *Buchnera* is an intermediate stage between a bacterium and an organelle! Genomes of *Buchnera* and its aphid host are interdependent (Pennisi 2009). For other reviews of aphid-bacterial symbioses, see Baumann, Moran, and Baumann (1997) and Baumann (2005, 2006).

17. Buchner (1965: 66, 68) was very critical of Jaroslav Peklo's work on insect-bacterial symbioses: "Inasmuch as Peklo's brief publications present no proof to support his statements, they cannot be regarded 'as serious contributions to symbiosis research.' . . . The 'symbionts' which Peklo (1946, 1949, 1951) still quite recently managed to find in the fatty tissue and ovaries of insects of all kinds . . . are obviously metabolic end products. . . ." See also note to p. 49.

18. Bacterial symbionts of Cicadidae (Insecta: Homoptera: Auchenorrhyncha) are well studied; Buchner (1965: 345–430) called this suborder of Homoptera a "fairyland of symbiosis"; see Moran, Tran, and Gerardo (2005).

19. Mycetomes in lice (Anoplura), notorious vectors of the typhus bacteria *Rickettsia prowazekii* and other pathogens, have been studied in detail (Buchner 1965: 482–507). Symbionts associated with midgut (stomach disc) bacteriocytes in human body and head lice belong to Gammaproteobacteria (Perotti et al. 2007); the recently proposed name is *Candidatus* Riesia pediculicola (Sasaki-Fukatsu et al. 2006).

20. Symbionts of ticks (Arachnida: Acari) residing in the ovarian tissue of females have been studied in detail (Rymaszewska 2007). Most recently, an intracellular alpha-proteobacterium, *Midichloria*, was discovered in *Ixodes ricinus*, which is even able to invade the mitochondria of the cells in which it resides (Sacchi et al. 2004; Lo et al. 2006; see Fig. II-4).

21. Symbioorgans (mycetomes) of leeches (Hirudinea; mainly Glossiphoniidae) are well studied; see Buchner (1965: 433–436) for a review of early work; modern information can be found in Siddall, Perkins, and Desser (2004). Bacterial symbionts belong to several groups, including Alphaproteobacteria close to Rhizobiaceae, Gammaproteobacteria, *Aeromonas*, and *Rickettsia* (Perkins, Budinoff, and Siddall 2005; Worthen, Gode, and Graf 2006).

22. The reference to fat-body study in moths (Insecta: Lepidoptera) (Pospelov 1922) is unclear. Probably Kozo-Polyansky is talking about a microsporidian fungus *Nosema* (phylum Microspora), now regarded as a common insect parasite. Kozo-Polyansky mentions similarity to the organism causing pebrine (nosematosis of silkmoth) that was famously studied by Pasteur. Buchner (1965: 69–70) mentions that *Nosema* was misinterpreted by Portier (1918) as a symbiont. Microsporidia are intensively studied as parasite models; their genomes are extraordinary among eukaryotes for their extreme reduction (Agnew et al. 2003; Slamovits et al. 2004).

23. The cockroach fat body is currently well accepted as a mycetome. The bacterial symbiont *Blattabacterium* (formerly *Bacillus*) *cuenoti* (Mercier 1906) Hollande and Favre 1931 is found in almost all cockroaches (Insecta: Blattoidea) (Buchner 1965: 516–527; Sacchi et al. 1996; Clark and Kambhampati 2004; Lo et al. 2007; see fig. III-7).

24. Bedbugs (Hemiptera: Cimicidae: *Cimex*) indeed have glands currently confirmed as mycetomes (bacteriomes); the eubacterial symbiont is *Tsukamurella paurometabolum*. See Goodfellow and Maldonado (2005) for systematics of this bacterial group.

25. The symbiotic nature of beetle luminescence is an erroneous assumption (Buchner 1965). No modern reviews assign luminescence (luciferin production) in fireflies (Coleoptera: Lampyridae) to symbionts. See further, however, on luminescent symbioorgans (bacteriomes) of squids and tunicates.

26. The bacterial symbiosis in the "storage kidney" of the land snail *Pomatias* (=*Cyclostoma*) *elegans* (Gastropoda: Prosobranchia: Pomatiasidae) is reviewed in Buchner (1965: 612–614), based only on the 1920s data. It is mentioned by Saffo (1992) as an example of heterotrophic bacterial endosymbionts in molluscs along with shipworms (Bilvalvia: Teredinidae), which harbor cellulolytic bacteria in their gills.

27. *Conferva*, mentioned by Darwin in his *Voyage of the* Beagle in his discovery of microbial mats, is no longer considered a genus of green algae. Rather it is a mixture of several genera or species.

28. Kozo-Polyansky's description of *Nephromyces*, based on Giard (1888), is hardly correct or complete. A very limited knowledge of *Nephromyces* existed for many decades (Buchner 1965). Recently, this symbiosis of Molgulidae (Tunicata) has been studied in detail (Saffo 1982, 1991; Saffo and Davis 1982; Saffo et al., in progress). *Nephromyces* proved to be not a chytrid but an apicomplexan protoctist (Saffo et al., in progress). Its infective stage is highly similar to sporozoites of Apicomplexa. In addition, it contains hereditarily transmitted intracellular beta-proteobacterial symbionts. *Nephromyces* (fig. III-11) was present in all studied adult *Molgula* species with 100 percent prevalence (Saffo and Nelson 1983; Saffo et al. in progress).

Giard (1888), who named *Nephromyces*, thought that all cells belonged to one taxon. However, already de Lacaze-Duthiers (1874) who first published information on this symbiosis, thought that it was a diverse community, but wondered whether the "confervoid" filaments might be gregarines (they proved to be mineral inclusions; Saffo and Lowenstam 1978; Saffo 1991). Hellman's (1913) drawings of the "spirochaetes" mentioned by Kozo-Polyansky are clearly the *Nephromyces* filaments.

Nephromyces (including its intracellular bacteria) possesses striking urate oxidase activity (Saffo 1991; Saffo and Nelson 1983). How the animal host benefits from this is still not clear (the removing "excreta" hypothesis is not likely, since the renal sac of lab-raised uninfected animals does not overfill) (Saffo et al. in progress). *Nephromyces* has been assumed to be a chytrid due to its "flagellated" (undulipodiated) stage, lacking in most Apicomplexa. Saffo et al. (in progress) assume that these undulipodia (two, rather than one, listed by Kozo-Polyansky after Giard) are related to sexual reproduction. Note that chytrids are no longer considered fungi; they belong to the protoctist class "Opisthokonta" and are ancestral to fungi. All fungi are amastigote (they lack undulipodia at all stages [Margulis and Chapman 2009]). We thank Dr. Mary Beth Saffo for her detailed comments on this amazing symbiosis.

29. Buchner (1965: 571–589) provides a review of early work on both pyrosomids and salpids with luminescent symbionts. For modern data, see Mackie and Bone (1978) and Bowlby, Widder, and Case (1990). See also Herring (2005) for a list of all bioluminescent organisms. Thomas Huxley observed, in 1849:

"I have just watched the moon set in all her glory, and looked at those lesser moons, the beautiful *Pyrosoma,* shining like white-hot cylinders in the water" (Huxley 1936).

30. The cephalopod-bacterial luminescent symbiosis (Mollusca: Cephalopoda) not only is indeed currently confirmed but became a great model for the study of bacterial-animal associations. Two families of squids (Loliginidae) and cuttlefish (Sepiolidae) harbor bacteriogenic light organs. The symbionts are *Vibrio fischeri.* A detailed review of early work was provided by Buchner (1965: 543–571). For modern information, see works of M. J. McFall-Ngai and collaborators on the Hawaiian sepiolid squid *Euprymna scolopes* (e.g., Nyholm et al. 2000 and McFall-Ngai 2000, 2002, 2008; see also fig. III-12) and others, e.g., Epel (1998); Barbieri et al. (2001); and Nishiguchi, Lopez, and Boletzky (2004).

31. A terminological confusion. The term "hematoblasts" (usually credited to Hayem [1877], but see Fatović-Ferenčić [1999] on the priority of Heitzmann [1872]]) refers to stem cells of blood located in bone marrow (also known as hemocytoblasts and pluripotent hematopoietic stem cells). Kozo-Polyansky is talking about platelets, or *thrombocytes,* for which Bizzozero (1882) introduced the term "blood plates" and documented their importance in blood coagulation (Cooper 2005). Platelets are not considered symbiotic by origin.

32. Schaudinn (1904) most likely observed not a symbiosis but a simple fungal infection (Buchner 1965: 52).

33. *Sambucus* enigmatic tubelike structures are now known to be coenocytic (multinuclear) tannin sacs (Evert 2006).

34. The enigmatic *"Cytorreictes"* (usually spelled *Cytoryctes*) refers to viruses (smallpox, cowpox) and was described in Protozoa as *Citoryctes* by Guarnieri (1892). *"Cytoryctes"* are now called "Guarnieri bodies," cytoplasmic inclusion bodies found in the epidermis and composed of viral particles and proteins. In old literature it is used as a generic Latin name—e.g., Solomon (1911) lists *Cytoryctes variolae* (smallpox) and *C. vaccinia* (cowpox). Paschen (1906) is credited with discovery of the viral nature of smallpox, but the issue of a protozoan parasite was not settled until the 1920s. See also Köhler (2001).

35. Some predictions by Kozo-Polyansky made on these pages have not been confirmed, while others have. The symbiotic nature of nidamental glands in squid, predicted by Pierantoni, was discovered (Buchner 1965: 73). At the same time, myriad other remarkable symbioses have been discovered since 1924; and of course the most spectacular prediction shared by Kozo-Polyansky with Merezhkovsky (1905, 1909a), Famintsyn (1907, 1912), Portier (1918), and Wallin (1922, 1927)—that mitochondria and plastids are symbiogenetic organelles—has been spectacularly confirmed (Margulis 1993).

IV. The Philosophy of Symbiogenesis

1. Ludwig H. Plate (1862–1937)—zoologist and geneticist, student and successor of Ernst Haeckel. One of the most influential evolutionary scientists in Europe in the first third of the twentieth century. He founded the journal *Archiv für Rassen- und Gesellschaftsbiologie* and supported the Nazi movement. See Levit and Hoßfeld (2006).

2. *Leaps in nature* (and society) and "transition from quality into quantity" are inverted-Hegel staple statements of Marxist philosophy, or "dialectical material-ism" (*"My dialectic method is not only different from the Hegelian, but is its direct opposite"* (Marx). Kozo-Polyansky was very fond of philosophy. As "dialectical materialism" became an official state doctrine and the only philosophy allowed in Russia, Kozo-Polyansky was actively introducing its approach and its lingo into biological science. As early as 1925, he wrote a book for high school teachers, *Dialectics of Biology*. At this time, Friedrich Engels's book *Dialectics of Nature* (1883), the mandatory Marxist text of later years, was not yet translated into Russian.

   Hegel wrote: "It is said, *natura non facit saltum* ["there are no leaps in nature"; compare to Kozo-Polyansky's 1921 statement, *natura facit saltum*, see our p. xxiv]; . . . But we have seen that the alterations of being in general are not only the transition of one magnitude into another, but a transition from quality into quantity and *vice versa,* a becoming-other which is an interruption of gradualness and the production of something qualitatively different from the reality which preceded it." (*Hegel's Science of Logic,* 1969, § 776) Contrast this with Karl Popper's *What Is Dialectic?* (1963: 316): "The whole development of dialectic should be a warning against the dangers inherent in philosophical system-building. It should remind us that philosophy should not be made a basis for any sort of scientific system and that philosophers should be much more modest in their claims. One task which they can fulfill quite usefully is the study of the critical methods of science."

3. *"Der Wiederspruch ist das Fortleitende,"* a popular Marxist statement, is probably quoted from G. Plekhanov's *The Development of the Monist View of History* (1895), where this quote appears twice.

4. St. George Jackson Mivart (1827–1890), the leading opponent of Darwin, author of *On the Genesis of Species* (1871).

5. Ludwig Gumplowicz (1838–1909), a Polish sociologist who worked in Austro-Hungary, often classified as a social Darwinist, author of *Geschichte der Staatstheorien* (History of theories of state) and *Der Rassenkampf* (Struggle of the races). For Gumplowicz, society was "nothing but an aggregate of groups which, in turn, completely determined the individual's thoughts, actions, and emotions" (Weiler 2007).

6.   *Two souls, alas, reside within my breast*
      And each withdraws from, and repels, its brother.

(Goethe, *Faust*, pt. 1, chap. 2, "Before the City Gate," translated by Bayard Taylor, 1870–1871).

V. History of Symbiogenesis Theory

1. The last sentence of the book, "And yet it does move!" ("Eppur si muove"), is ascribed to Galileo.

References to Kozo-Polyansky's Text

1. This green alga, *Cephaleuros virescens* (Chlorophyta: Trentepohliales) is known as algal leaf spot; grows in close association with fungi, forming lichens, often in the tropics. It has been now reported from 218 genera and 62 families of vascular plants in the Gulf of Mexico coast in the United States, and in 119 genera from Brazil. See Holcomb (1986), López-Bautista, Waters, and Chapman (2002).

Commentary References

Abella, C. A., X. P. Cristina, A. Martinez, I. Pibernat, and X. Vila. 1998. "Two new motile phototrophic consortia: *'Chlorochromatium lunatum'* and *'Pelochromatium selenoides'.*" *Archives of Microbiology* 169: 452–459.

Adams, D. G. 2000. "Symbiotic interactions." In *Ecology of Cyanobacteria: Their Diversity in Time and Space*, ed. B. Whitton and M. Potts, pp. 523–561. Dordrecht, The Netherlands: Kluwer Academic.

Adams, D. G., B. Bergman, S. A. Nierzwicki-Bauer, A. N. Rai, and A. Schüßler. 2006. "Cyanobacterial-plant symbioses". In *The Prokaryotes. A Handbook on the Biology of Bacteria, third edition. Volume 1: Symbiotic Associations, Biotechnology, Applied Microbiology*, ed. M. Dworkin, S. Falkow, E. Rosenberg, K.-H. Schleifer, and E. Stackebrandt, pp. 331–363. New York: Springer.

Adams, D. G., and P. S. Duggan. 2008. "Cyanobacteria-bryophyte symbioses." *Journal of Experimental Botany* 59: 1047–1058.

Agnew, P., J. J. Becnel, D. Ebert, and Y. Michalakis. 2003. "Symbiosis of Microsporidia and insects." In *Insect Symbiosis,* ed. K. Bourtzis and T. A. Miller, pp. 145–164. Boca Raton, Fl.: CRC.

Ahmadjian, V. 1993. *The Lichen Symbiosis.* New York: John Wiley.

Andersson, J. O. 2000. "Evolutionary genomics: Is *Buchnera* a bacterium or an organelle?" *Current Biology* 10: R866–868.

Andersson, S. G., O. Karlberg, B. Cänback, and C. C. Kurland. 2003. "On the origin of mitochondria: A genomics perspective." *Philosophical Transactions of the Royal Society London B, Biological Sciences* 358: 165–177.

Baker, J. A., B. Entsch, and D. B. McKay. 2003. "The cyanobiont in an *Azolla* fern is neither *Anabaena* nor *Nostoc.*" *FEMS Microbiology Letters* 229: 43–47.

Barbieri, E., B. J. Paster, D. Hughes, L. Zurek, D. P. Moser, A. Teske, and M. L. Sogin. 2001. "Phylogenetic characterization of epibiotic bacteria in the accessory nidamental gland and egg capsules of the squid *Loligo pealei* (Cephalopoda: Loliginidae)." *Environmental Microbiology* 3: 151–167.

Baumann, P. 2005. "Biology of bacteriocyte-associated endosymbionts of plant sap-sucking insects." *Annual Revew of Microbiology* 59: 155–189.

———. 2006. "Diversity of prokaryote-insect associations within the Sternorrhyncha (psyllids, whiteflies, aphids, mealybugs)." In *Insect Symbiosis,* ed. K. Bourtzis and T. A. Miller, pp. 2:1–24. Boca Raton, Fl.: CRC.

Baumann, P., L. Baumann, M. A. Clark, and M. L. Thao. 1998. *"Buchnera aphidicola:* The endosymbiont of aphids." *ASM News* 64: 203–209.

Baumann, P., N. A. Moran, and L. Baumann. 1997. "The evolution and genetics of aphid endosymbionts." *BioScience* 47: 12–20.

Benson, J., and L. Margulis. 2002. "The *Gunnera manicata—Nostoc* symbiosis: Is the red stipulate tissue symbiogenetic?" *Biology and Environment: Proceedings of the Royal Irish Academy* 102b: 45–48.

Bergman, B. 2002. "The *Nostoc-Gunnera* symbiosis." In *Cyanobacteria in Symbiosis,* ed. A. N. Rai, B. Bergman, and U. Rasmussen, pp. 207–232. Dordrecht, The Netherlands: Kluwer Academic.

Bergman, B., and B. Osborne. 2002. "The *Gunnera-Nostoc* symbiosis." In *Commentaries on Cyanobacterial Symbioses,* ed. B. Osborne, pp. 35–39. Dublin: Royal Irish Academy.

Bizzozero, G. 1882. "Ueber einer neuen Formbestandtheil des Blutes und dessen Rolle bei der Thrombose und der Blutgerinnung." *Archiv für pathologische Anatomie und Physiologie und für klinische Medizin* 90: 261–332.

Bourtzis, K., and T. A. Miller, eds. 2003. *Insect Symbiosis.* Boca Raton, Fl.: CRC.

———, eds. 2006. *Insect Symbiosis,* vol. 2. Boca Raton, Fl.: CRC.

Bowlby, M. R., E. A. Widder, and J. F. Case. 1990. "Patterns of stimulated bioluminescence in two pyrosomes (Tunicata: Pyrosomatidae)." *Biological Bulletin* 179: 340–350.

Brodo, I. M., S. D. Sharnoff, and S. Sharnoff. 2001. *Lichens of North America.* New Haven: Yale University Press.

Buchner, P. 1965. *Endosymbiosis of Animals with Plant Microorganisms.* New York: Interscience.

Chakrabarty, D., A. K. Mandal, and S. K. Datta. 2000. "Retrieval of new coloured chrysanthemum through organogenesis from sectorial chimera." *Current Science* 78: 1060–1061.

Chapman, M., M. F. Dolan, and L. Margulis. 2000. "Centrioles and kinetosomes: Form, function, and evolution." *Quarterly Review of Biology* 75: 409–429.

Chatton, E. 1925. *"Pansporella perplexa* amoebien à spores protégées parasite des Daphnies." *Annales des Sciences Naturelles Zoologie, Série 10,* 8: 5–84.

Chigira, M. 1997. "Transplantation and chimera as extended self." *Medical Hypotheses* 48: 63–69.

Clark, J. W., and S. Kambhampati. 2004. "Phylogenetic analysis of *Blattabacterium,* endosymbiotic bacteria from the wood roach, *Cryptocercus* (Blattodea: Cryptocercidae), including a description of three new species." *Molecular Phylogenetics and Evolution* 26: 82–88.

Claus, G. 1961. "Observations on *Cyanotheca longipes* Pascher." *Plant Systematics and Evolution* 108: 286–299.

Clay, K., and C. Schardl. 2002. "Evolutionary origins and ecological consequences of endophyte symbiosis with grasses." *American Naturalist* 160, suppl. 4: S99–S127.

Cooper, B. 2005. "Osler's role in defining the third corpuscle, or 'blood plates.'" *Proceedings of Baylor University Medical Center* 18: 376–378.

Cox, P. A., S. A. Banack, S. J. Murch, U. Rasmussen, G. Tien, R. R. Bidigare, J. S. Metcalf, L. F. Morrison, G. A. Codd, and B. Bergman. 2005. "Diverse taxa of cyanobacteria produce β-N-methylamino-L-alanine, a neurotoxic amino acid." *Proceedings of the National Academy of Sciences of the USA* 102: 5074–5078.

Dearnaley, J. D. 2007. "Further advances in orchid mycorrhizal research." *Mycorrhiza* 17: 475–486.

Degnan, P. H., A. B. Lazarus, C. D. Brock, and J. J. Wernegreen. 2004. "Host-symbiont stability and fast evolutionary rates in an ant-bacterium association: Cospeciation of *Camponotus* species and their endosymbionts, *Candidatus Blochmannia*." *Systematic Biology* 53: 95–110.

de Souza, D. J., A. Bézier, D. Depoix, J.-M. Drezen, and A. Lenoir. 2009. "*Blochmannia* endosymbionts improve colony growth and immune defence in the ant *Camponotus fellah*." *BMC Microbiology* 9: 29.

Dolan, M. F., H. Melnitsky, L. Margulis, and R. Kolnicki. 2002. "Motility proteins and the origin of the nucleus." *Anatomical Record* 268: 290–301.

Douglas, A. E. 1989. "Mycetocyte symbiosis in insects." *Biological Reviews of the Cambridge Philosophical Society* 64: 409–434.

———. 1998. "Nutritional interactions in insect-microbial symbioses: Aphids and their symbiotic bacteria *Buchnera*." *Annual Review of Entomology* 43: 17–37.

———. 2003. "*Buchnera* bacteria and other symbionts of aphids." In *Insect Symbiosis.*, ed. K. Bourtzis and T. A. Miller, pp. 23–38. Boca Raton, Fl.: CRC.

Eissenstat, D. M. 1990. "A comparison of phosphorus and nitrogen transfer between plants of different phosphorus status." *Oecologia* (Berlin) 82: 342–347.

Elmerich, C., and W. E. Newton, eds. 2003. *Associative Nitrogen-Fixing Bacteria and Cyanobacterial Associations*. Dordrecht, The Netherlands: Kluwer Academic.

Enderlein, G. 1925. *Bakterien-Cyclogenie. Prolegomena zu Untersuchungen über Bau, geschlechtliche und ungeschlechtliche Fortpflanzung und Entwicklung der Bakterien*. Berlin and Leipzig: Walter de Gruyter.

Epel, D. 1998. "Bacterial symbionts colonize the accessory nidamental gland of the squid *Loligo opalescens* via horizontal transmission." *Biological Bulletin* 194: 36–43.

Evert, R. F. 2006. *Esau's Plant Anatomy*. 3rd. ed. Hoboken, N.J.: John Wiley & Sons.

Fatović-Ferenčić, S. 1999. "The description of the hematoblast by the dermatopathologist Carl Heitzmann in 1872." *Journal of Investigative Dermatology* 113: 861–862.

Finlay, R. D. 2008. "Ecological aspects of mycorrhizal symbiosis: With special emphasis on the functional diversity of interactions involving the extraradical mycelium." *Journal of Experimental Botany* 59: 1115–1126.

Geus, A., and E. Höxtermann, eds. 2007. *Evolution durch Kooperation und Integration. Acta Biohistorica* 11, 751 S., 111 Abb. Marburg an der Lahn: Basilisken-Presse.

Gil, R., B. Sabater-Muñoz, A. Latorre, F. J. Silva, and A. Moya. 2002. "Extreme genome reduction in *Buchnera* spp.: Toward the minimal genome needed for symbiotic life." *Proceedings of the National Academy of Sciences of the USA* 99: 4454–4458.

Goodfellow, M., and L. A. Maldonado. 2006. "The Families Dietziaceae, Gordoniaceae, Nocardiaceae and Tsukamurellaceae." In *The Prokaryotes. A Handbook on the Biology of Bacteria, third edition. Volume 3, Archaea and Bacteria. Firmicutes, Actinomycetes*, ed. M. Dworkin, S. Falkow, E. Rosenberg, K.-H. Schleifer, and E. Stackebrandt, pp. 843–888. New York: Springer.

Guarnieri, G. 1892. "Ricerche sulla patogenesi ed etiologia dell'infezione vaccinica e valolosa." *Archivio per le scienze mediche* 16: 403–423.

Guerrero, R., and M. Belanguer. 2010. "The Symbioma." *Int. Journal of Microbiology*. In press.

Guerrero, R., and L. Margulis. 2010. "Symbiomas: The bacterial cell as the unit of life." *Proceedings of the National Academy of Sciences of the USA*. In preparation.

(Hegel, G. W. F.) *Hegel's Science of Logic*. 1969. Translated by A. V. Miller. Foreword by J. N. Findlay. London: G. Allen & Unwin.

Heitzmann, C. 1872. "Studien am Knochen und Knorpel (ueber Blutbildung im entzündeten Knochen)." *Medizinische Jahrbücher* 1872: 339–366.

Herring, P. J. 2005. "Systematic distribution of bioluminescence in living organisms." *Journal of Bioluminescence and Chemiluminescence* 1: 147–163.

Holcomb, G. E. 1986. "Hosts of the parasitic alga *Cephaleuros virescens* in Louisiana and new host records for the continental United States." *Plant Disease* 70: 1080–1083.

Honegger, R. 1991. "Fungal evolution: Symbiosis and morphogenesis." In *Symbiosis as a Source of Evolutionary Innovation: Speciation and Morphogenesis*, ed. L. Margulis and R. Fester, pp. 319–340. Cambridge: MIT Press.

Huxley, J. 1949. *Soviet Genetics and World Science: Lysenko and the Meaning of Heredity*. London: Chatton and Windus.

Huxley, T. H. 1936. *T. H. Huxley's Diary of the Voyage of H.M.S. Rattlesnake*. Garden City, N.Y.: Doubleday.

Imhoff, J. F. 2003. "Phylogenetic taxonomy of the family Chlorobiaceae on the basis of 16S rRNA and *fmo* (Fenna-Matthews-Olson protein) gene sequences." *International Journal of Systematic and Evolutionary Microbiology* 53: 941–951.

Izumi, Y., K. Ono, M. Takamiya, and K. Fukui. 1993. "Chloroplast division in cultured cells of the hornwort *Anthoceros punctatus.*" *Journal of Plant Research* 106: 319–325.

Johansson, C., and B. Bergman. 1994. "Reconstitution of the *Gunnera manicata* Linden symbiosis: Cyanobacterial specificity." *New Phytologist* 126: 643–652.

Jolivet, P. 1998. *Interrelationship between Insects and Plants.* Boca Raton, Fl.: CRC Press.

Kanzler, B. E. M., K. R. Pfannes, K. Vogl, and J. Overmann. 2005. "Molecular characterization of the nonphotosynthetic partner bacterium in the consortium *'Chlorochromatium aggregatum.'*" *Applied and Environmental Microbiology* 71: 7434–7441.

Karandashov, V., and M. Bucher. 2005. "Symbiotic phosphate transport in arbuscular mycorrhizas." *Trends in Plant Science* 10: 22–29.

Khakhina, L. N. 1992. *Concepts of Symbiogenesis: A Historical and Critical Study of the Research of Russian Botanists.* Edited by L. Margulis and M. McMenamin. New Haven: Yale University Press.

Köhler, W. 2001. "Zentralblatt für Bakteriologie—100 years ago: Protozoa as causative agents of smallpox, Or: *Cytoryctes* and no end." *International Journal of Medical Microbiology* 291: 191–195.

Kolchinsky, E. I. 1991. "The 'union' of philosophy and biology that did not happen (1920s–1930s)" [in Russian]. In *Repressirovannaya nauka* (The repressed science), pp. 34–70. Leningrad: Nauka.

———. 1999. *V poiskakh sovetskogo "soyuza" filosofii i biologii (diskussii i repressii v 20kh–nachale 30kh gg.)* (In Search of a Soviet "Union" of Philosophy and Biology [Discussions and Repressions of 1920s—Early 1930s]). St. Petersburg: Dm. Bulanin.

———. 2001. "Darwinismus und Marxismus in der Epoche des frühen Stalinismus." In *Darwinismus und/als Ideologie,* ed. U. Hoßfeld and R. Brömer, pp. 157–166. Berlin: Verlag für Wissenschaft und Bildung.

Komárek J., and T. Hauer. 2004. "CyanoDB.cz—On-line database of cyanobacterial genera." http://www.cyanodb.cz. Accessed November 2009.

Kozo-Polyansky, B. M. 1922a. *Final evolyutsii* (The goal of evolution). Moscow: Burevestnik.

———. 1922b. *Vvedenie v filogeneticheskuyu sistematiku vysshikh rastenii* (An introduction into phylogenetic systematics of higher plants). Voronezh: Priroda i kul'tura.

———. 1924. *Novyi printsip biologii: ocherk teorii simbiogeneza* (The New Principle of Biology: An Essay of the Theory of Symbiogenesis). Moscow-Leningrad: Puchina.

———. 1925. *Dialektika v biologii* (Dialectics in biology). Rostov-na-Donu: Krasnodar.

———. 1951. "Against idealism in plant morphology" [in Russian]. *Botanicheskii zhurnal* (Botanical journal) (Moscow) 36: 115–128.

———. 1965. *Kurs sistematiki vysshikh rastenii* (A course in systematics of higher plants). Voronezh: Voronezhskii gos. universitet.

Lacaze-Duthiers, F. J. H. de. 1874. "Histoire des ascidies simples des côtes de France I." *Archives de zoologie expérimentale et générale* 3: 119–656.

Latorre, A., and A. Moya. 2006. "Comparative genomics in *Buchnera aphidicola,* primary endosymbiont of aphids." In *Insect Symbiosis,* ed. K. Bourtzis and T. A. Miller, pp. 2:157–174. Boca Raton, Fl.: CRC.

Leigh, J., M. A. Hodge, and A. H. Fitter. 2009. "Arbuscular mycorrhizal fungi can transfer substantial amounts of nitrogen to their host plant from organic material." *New Phytologist* 181: 199–207.

Levit, G. S., and U. Hoßfeld. 2005. "Die Nomogenese: Eine Evolutionstheorie jenseits des Darwinismus und Lamarckismus." *Verhandlungen zur Geschichte und Theorie der Biologie* 11: 367–388.

———. 2006. "The forgotten "old-Darwinian" synthesis: The evolutionary theory of Ludwig H. Plate (1862–1937)." *NTM International Journal of History and Ethics of Natural Sciences, Technology and Medicine* 14: 9–25.

Levit, G. S., U. Hoßfeld, and L. Ollson. 2006. "From the 'Modern Synthesis' to cybernetics: Ivan Ivanovich Schmalhausen (1884–1963) and his research program for a synthesis of evolutionary and developmental biology." *Journal of Experimental Zoology B, Molecular and Developmental Evolution* 306: 89–106.

Lo, N., T. Beninati, L. Sacchi, and C. Bandi. 2006. "An alpha-proteobacterium invades the mitochondria of the tick *Ixodes ricinus.*" In *Insect Symbiosis,* ed. K. Bourtzis and T. A. Miller, pp. 2:25–37. Boca Raton, Fl.: CRC.

Lo, N., T. Beninati, F. Stone, J. Walker, and L. Sacchi. 2007. "Cockroaches that lack *Blattabacterium* endosymbionts: The phylogenetically divergent genus *Nocticola.*" *Biology Letters* 3: 327–330.

López-Bautista J. M., D. A. Waters, and R. L. Chapman. 2002. "The Trentepohliales revisited." *Constancea* 83: 1–21. http://ucjeps.berkeley.edu/constancea/83/. Accessed November 2009.

Mackie, G. O., and Q. Bone. 1978. "Luminescence and associated effector activity in *Pyrosoma* (Tunicata: Pyrosomida)." *Proceedings of the Royal Society of London, Series B* 202: 483–495.

Marcotrigiano, M. 1997. "Chimeras and variegation: Patterns of deceit." *Horticultural Science* 32: 773–784.

Margulis, L. 1993. *Symbiosis in Cell Evolution.* 2nd ed. New York: Freeman.

Margulis, L., and M. Chapman. 1998. "Endosymbioses: Cyclical and permanent in evolution." *Trends in Microbiology* 6: 342–345.

———, eds. 2009. *Kingdoms and Domains: Illustrated Guide to the Phyla of Life.* New York: Academic Press, Elsevier.

Margulis, L., M. Chapman, R. Guerrero, and J. Hall. 2006. "The last eukaryotic common ancestor (LECA): Acquisition of cytoskeletal motility from aerotolerant spirochetes in the Proterozoic Eon." *Proceedings of the National Academy of Sciences of the USA* 103: 13080–13085.

Margulis, L., M. Dolan, and R. Guerrero. 2000. "The chimeric eukaryote: Origin of the nucleus from the karyomastigont in amitochondriate protists." *Proceedings of the National Academy of Sciences of the USA* 97: 6954–6959.

Margulis, L., M. Enzien, and H. I. McKhann. 1990. "Revival of Dobell's 'chromidia' hypothesis: Chromatin bodies in the amoebomastigote *Paratetramitus jugosus.*" *Biological Bulletin* 178: 300–304.

Margulis, L., and R. Fester, eds. 1991. *Symbiosis as a Source of Evolutionary Innovation: Speciation and Morphogenesis.* Cambridge: MIT Press.

Margulis, L., and M. McMenamin. 1990. "Marriage of convenience: The motility of the modern cell may reflect an ancient symbiotic union." *The Sciences* 30(5):30–37.

Margulis, L., and D. Sagan. 2002. *Acquiring Genomes: A Theory of the Origins of Species.* Amherst, Mass.: Perseus Books Group.

McFall-Ngai, M. J. 2000. "Negotiations between animals and bacteria: The 'diplomacy' of the squid-vibrio symbiosis." *Comparative and Biochemical Physiology, Part A—Molecular and Integrative Physiology* 126: 471–480.

———. 2002. "Unseen forces: The influence of bacteria on animal development." *Developmental Biology* 242: 1–14.

———. 2008. "Host-microbe symbiosis: The squid-*Vibrio* association–A naturally occurring, experimental model of animal/bacterial partnerships." *Advances in Experimental Medicine and Biology* 635: 102–112.

Medvedev, Z. A. 1969. *Rise and Fall of T. D. Lysenko.* New York: Columbia University Press.

Moran, N. A., and P. Baumann. 2000. "Bacterial endosymbionts in animals." *Current Opinions in Microbiology* 3: 270–275.

Moran, N. A., and P. H. Degnan. 2006. "Functional genomics of *Buchnera* and the ecology of aphid hosts." *Molecular Ecology* 15: 1251–1261.

Moran, N. A., P. Tran, and N. M. Gerardo. 2005. "Symbiosis and insect diversification: An ancient symbiont of sap-feeding insects from the bacterial phylum *Bacteroidetes.*" *Applied and Environmental Microbiology* 71: 8802–8810.

Mujer, C. V., D. L. Andrews, J. R. Manhart, S. K. Pierce, and M. E. Rumpho. 1996. "Chloroplast genes are expressed during intracellular symbiotic association of *Vaucheria litorea* plastids with the sea slug *Elysia chlorotica.*" *Cell Biology* 93: 12333–12338.

Nakahashi, C. D., K. Frole, and L. Sack. 2005. "Bacterial leaf nodule symbiosis in *Ardisia* (Myrsinaceae): Does it contribute to seedling growth capacity?" *Plant Biology* 7: 495–500.

Nilsson, P., B. Bergman, and U. Rasmussen. 2000. "Cyanobacterial diversity in geographically related and distant host plants of the genus *Gunnera.*" *Archives of Microbiology* 173: 97–102.

Nishiguchi, M. K., J. E. Lopez, and S. Boletzky. 2004. "Enlightenment of old ideas from new investigations: More questions regarding the evolution of bacteriogenic light organs in squids." *Evolution and Development* 6: 41–49.

Noda, H., and K. Kodama. 1996. "Phylogenetic position of yeastlike endosymbionts of anobiid beetles." *Applied and Environmental Microbiology* 62: 162–167.

Nyholm, S. V., E. V. Stabb, E. G. Ruby, and M. J. McFall-Ngai. 2000. "Establishment of an animal–bacterial association: Recruiting symbiotic vibrios from the environment." *Comparative Biochemistry and Physiology, Part A—Molecular and Integrative Physiology* 126: 471–480.

Okamoto, N., and I. Inouye. 2006. "*Hatena arenicola* gen. et sp. nov., a katablepharid undergoing probable plastid acquisition." *Protist* 157: 401–419.

Overmann, J. 2006. "Symbiosis between non-related bacteria in phototrophic consortia." *Progress in Molecular and Subcellular Biology* 41: 21–37.

Overmann, J., and K. Schubert. 2002. "Phototrophic consortia: Model systems for symbiotic interrelations between prokaryotes." *Archives of Microbiology* 177: 201–208.

Paschen, E. 1906. "Was wissen wir über den Vakzineerreger?" *Munchener medizinische Wochenschrift* 53: 2391–2395.

Pennisi, E. 2009. "The Biology of Genomes, 5–9 May 2009, Cold Spring Harbor, New York. The bug and the bacterium: Interdependent genomes." *Science* 324 (5932): 1253.

Perkins, S. L., R. B. Budinoff, and M. E. Siddall. 2005. "New Gammaproteobacteria associated with blood-feeding leeches and a broad phylogenetic analysis of leech endosymbionts." *Applied and Environmental Microbiology* 71: 5219–5224.

Perotti, M. A., J. M. Allen, D. L. Reed, and H. R. Braig. 2007. "Host-symbiont interactions of the primary endosymbiont of human head and body lice." *FASEB Journal* 21: 1058–1066.

Plekhanov, G. V. 1895. *K voprosu o razvitii monisticheskogo vzglyada na istoriyu* (The Development of the Monist View of History). St. Petersburg.

Popper, K. 1962. *Conjectures and Refutations: The Growth of Scientific Knowledge.* New York: Basic Books.

Powell, M. J. 1979. "What are the chromidia of *Polyphagus euglenae?*" *American Journal of Botany* 66: 1173–1180.

Rai, A. N., B. Bergman, and U. Rasmussen, eds. 2002. *Cyanobacteria in Symbiosis.* Dordrecht, The Netherlands: Kluwer Academic.

Rai, A. N., E. Söderbäck, and B. Bergman. 2000. "Cyanobacterium-plant symbioses." *New Phytologist* 147: 449–481.

Read, D. J. 2005. "Mycorrhizas in ecosystems." *Cellular and Molecular Life Sciences* 47: 376–391.

Reyes-Prieto, A., A. P. Weber, and D. Bhattacharya. 2007. "The origin and establishment of the plastid in algae and plants." *Annual Review of Genetics* 41: 147–168.

Rumpho, M. E., J. M. Worful, J. Lee, K. Kannan, M. S. Tyler, D. Bhattacharya, A. Moustafa, and J. R. Manhart. 2008. "Horizontal gene transfer of the algal nuclear gene *psbO* to the photosynthetic sea slug *Elysia chlorotica.*" *Proceedings of the National Academy of Sciences of the USA* 105: 17867–17871.

Rymaszewska, A. 2007. "Symbiotic bacteria in oocyte and ovarian cell mitochondria of the tick *Ixodes ricinus:* Biology and phylogenetic position." *Parasitology Research* 100: 917–920.

Sacchi, L., E. Bigliardi, S. Corona, T. Beninati, N. Lo, and A. Franceschi. 2004. "A symbiont of the tick *Ixodes ricinus* invades and consumes mitochondria in a mode similar to that of the parasitic bacterium *Bdellovibrio bacteriovorus.*" *Tissue and Cell* 36: 43–53.

Sacchi, L., S. Corona, A. Grigolo, U. Laudani, M. G. Selmi, and E. Bigliardi. 1996. "The fate of the endocytobionts of *Blattella germanica* L. (Blattaria: Blattellidae) and *Periplaneta americana* (Blattaria: Blattidae) during embryo development." *Italian Journal of Zoology* 63: 1–11.

Saffo, M. B. 1982. "Distribution of the endosymbiont *Nephromyces* Giard within the ascidian family Molgulidae." *Biological Bulletin* 162: 95–104.

———. 1990. "Symbiosis within a symbiosis: Intracellular bacteria in the endosymbiotic protist *Nephromyces.*" *Marine Biology* 107: 291–296.

———. 1991. "Symbiogenesis and the evolution of mutualism: Lessons from the *Nephromyces*-bacterial-molgulid symbiosis." In *Symbiosis as a Source of Evolutionary Innovation: Speciation and Morphogenesis*, ed. L. Margulis and R. Fester, pp. 410–429. Cambridge: MIT Press.

———. 1992. "Invertebrates in endosymbiotic associations." *American Zoologist* 32: 557–565.

Saffo, M. B., and W. L. Davis. 1982. "Modes of infection of the ascidian *Molgula manhattensis* by its endosymbiont *Nephromyces* Giard." *Biological Bulletin* 162: 105–112.

Saffo, M. B., and H. A. Lowenstam. 1978. "Calcareous deposits in the renal sac of a molgulid tunicate." *Science* 200: 1166–1168.

Saffo, M. B., and R. Nelson. 1983. "The cells of *Nephromyces:* Developmental stages of a single life cycle." *Canadian Journal of Botany* 61: 3230–3239.

Sapp, J. 1994. *Evolution by Association: A History of Symbiosis.* New York: Oxford University Press.

———. 1998. "Freewheeling centrioles." *History and Philosophy of Life Sciences* 20: 255–290.

———. 2002. "Paul Buchner (1886–1978) and hereditary symbiosis in insects." *International Microbiology* 5: 145–150.

———. 2003. *Genesis: The Evolution of Biology.* New York: Oxford University Press.

———. 2005. "The prokaryote-eukaryote dichotomy: Meanings and mythology." *Microbiology and Molecular Biology Reviews* 69: 292–305.

Sapp, J., F. Carrapico, and M. Zolotonosov. 2002. "Symbiogenesis: The hidden face of Constantin Merezhkowsky." *History and Philosophy of Life Sciences* 24: 413–440.

Sasaki-Fukatsu, K., R. Koga, N. Nikoh, K. Yoshizawa, S. Kasai, M. Mihara, M. Kobayashi, T. Tomita, and T. Fukatsu. 2006. "Symbiotic bacteria associated with stomach disc of human lice." *Applied and Environmental Microbiology* 72: 7349–7352.

Schardl, C. L., A. Leuchtmann, and M. J. Spiering. 2004. "Symbioses of grasses with seedborne fungal endophytes." *Annual Review of Plant Biology* 55: 315–340.

Schoenwaelder, M. E. A. 2002. "The occurrence and cellular significance of physodes in brown algae." *Phycologia* 41: 125–139.

Schoenwaelder, M. E. A., and Clayton, M. N. 1999. "The role of cytoskeleton in brown algal physode movement." *European Journal of Phycology* 34: 223–229.

Searcy, D. G. 2003. "Metabolic integration during the evolutionary origin of mitochondria." *Cell Research* 13: 229–238.

Shigenobu S., H. Watanabe, M. Hattori, Y. Sakaki, and H. Ishikawa. 2000. "Genome sequence of the endocellular bacterial symbiont of aphids *Buchnera* sp. APS." *Nature* 407: 81–86.

Siddall, M. E., S. L. Perkins, and S. S. Desser. 2004. "Leech mycetome endosymbionts are a new lineage of alphaproteobacteria related to the Rhizobiaceae." *Molecular Phylogenetics and Evolution* 30: 178–186.

Slamovits, C. H., N. M. Fast, J. S. Law, and P. J. Keeling. 2004. "Genome compaction and stability in microsporidian intracellular parasites." *Current Biology* 14: 891–896.

Sleigh, M. A. 1991. *Protozoa and Other Protists*. 2nd ed. London: Edward Arnold.

Solomon, H. C. 1911. "The etiology of trachoma." *Transactions of the American Microscopical Society* 30: 41–55.

Strack, D., T. Fester, W. Schliemann, and M. H. Walter. 2003. "Arbuscular mycorrhiza: Biological, chemical, and molecular aspects." *Journal of Chemical Ecology* 29: 1955–1979.

Takhtajan, A. L. 1973. "Four kingdoms of the organic world" [in Russian]. *Priroda* (Nature) (Moscow) 2: 22–32.

———. 1983. "Macroevolutionary processes in the history of the plant world" [in Russian]. *Botanicheskii zhurnal* (Botanical journal) (Moscow) 68: 1593–1603.

Taylor, T. N., and M. Krings. 2005. "Fossil microorganisms and land plants: Associations and interactions." *Symbiosis* 40: 119–135.

Thao, M. L., P. J. Gullan, and P. Baumann. 2002. "Secondary (gamma-Proteobacteria) endosymbionts infect the primary (beta-Proteobacteria) endosymbionts of mealybugs multiple times and coevolve with their hosts." *Applied and Environmental Microbiology* 68: 3190–3197.

Thomas, L. 1978. *The Lives of a Cell: Notes of a Biology Watcher*. New York: Penguin.

Trautmann, D. A., and R. Hinde. 2002. "Sponge/algal symbioses; a diversity of associations." In *Symbiosis: Mechanisms and Model Systems*, ed. J. Seckbach, pp. 521–537. Dordrecht, The Netherlands: Kluwer Academic.

Van Oevelen, S., R. DeWachter, P. Vandamme, E. Robbrecht, and E. Prinsen. 2002. "Identification of the bacterial endosymbionts in leaf galls of *Psychotria* (Rubiaceae, Angiosperms) and proposal of 'Candidatus *Burkholderia kirkii*' sp. nov." *International Journal of Systematic and Evolutionary Microbiology* 52: 2023–2027.

Van Oppen, M. J., J. C. Mieog, C. A. Sánchez, and K. E. Fabricius. 2005. "Diversity of algal endosymbionts (zooxanthellae) in octocorals: The roles of geography and host relationships." *Molecular Ecology* 14: 2403–2417.

Venn, A. A., J. E. Loram, and A. E. Douglas. 2008. "Photosynthetic symbioses in animals." *Journal of Experimental Botany* 59: 1069–1080.

Vogl, K., J. Glaeser, K. R. Pfannes, G. Wanner, and J. Overmann. 2006. "*Chlorobium chlorochromatii* sp. nov., a symbiotic green sulfur bacterium isolated from the phototrophic consortium *'Chlorochromatium aggregatum.'*" *Archives of Microbiology* 185: 363–372.

Vogl, K., R. Wenter, M. Dreßen, M. Schlickenrieder, M. Plöscher, L. Eichacker, and J. Overmann. 2008. "Identification and analysis of four candidate symbiosis genes from *'Chlorochromatium aggregatum,'* a highly developed bacterial symbiosis." *Environmental Microbiology* 10: 2842–2856.

Wallin, I. E. 1927. *Symbionticism and the Origin of Species.* Baltimore: Williams and Wilkins.

Wanner, G., K. Vogl, and J. Overmann. 2008. "Ultrastructural characterization of the prokaryotic symbiosis in consortium *'Chlorochromatium aggregatum.'*" *Journal of Bacteriology* 190: 3721–3730.

Weiler, B. 2007. "Ludwig Gumplowicz." In *Blackwell Encyclopedia of Sociology Online*, ed. G. Ritzer, pp. 2038–2039. http://www.sociologyencyclopedia.com.

Wier, A. M., L. Sacchi, M. F. Dolan, C. Bandi, N. Lo, J. MacAllister, and L. Margulis. 2010. "Spirochete attachment ultrastructure: Implications for the origin and evolution of cilia." *Biological Bulletin* (in press).

Worthen, P. L., C. J. Gode, and J. Graf. 2006. "Culture-independent characterization of the digestive-tract microbiota of the medicinal leech reveals a tripartite symbiosis." *Applied and Environmental Microbiology* 72: 4775–4781.

Modern Classification of Life

Most inclusive "higher" taxa (Margulis and Chapman 2010)

SUPERKINGDOM PROKARYA

Origins not by symbiogenesis

KINGDOM PROKARYOTAE

(Bacteria, Monera, Prokarya)

SUBKINGDOM (DOMAIN) ARCHAEA

Division Mendosicutes (deficient-walled archaebacteria)

Phylum B-1 **Euryarchaeota** methanogens and halophils
Phylum B-2 **Crenarchaeota** thermoacidophils

SUBKINGDOM (DOMAIN) EUBACTERIA

Division Gracilicutes (Gram-negative bacteria)

Phylum B-3 **Proteobacteria** purple bacteria: phototrophs, heterotrophs
Phylum B-4 **Spirochaetae** helical motile heterotrophs, periplasmic flagella
Phylum B-5 **Bacteroides-Saprospirae** gliding fermenters, heterotrophs
Phylum B-6 **Cyanobacteria** oxygenic photoautotrophs
Phylum B-7 **Chloroflexa** gliding nonsulfur oxygen-tolerant photoautotrophs
Phylum B-8 **Chlorobia** sulfur oxygen-intolerant photoautotrophs

Division Tenericutes (wall-less eubacteria)

Phylum B-9 **Aphragmabacteria** no cell walls

Division Firmicutes (Gram-positive and protein-walled bacteria)

Phylum B-10 **Endospora** low-G+C endospore-forming Gram-positives and relatives

Phylum B-11 **Pirellulae** proteinaceous wall-formers and relatives

Phylum B-12 **Actinobacteria** fungoid multicellular Gram-positives and relatives

Phylum B-13 **Deinococci** radiation- or heat-resistant Gram-positives

Phylum B-14 **Thermotogae** thermophilic fermenters

SUPERKINGDOM EUKARYA

Origins by symbiogenesis

KINGDOM PROTOCTISTA

Four modes: phyla whose members
I. lack both undulipodia and meiotic sex
II. lack undulipodia but have meiotic sexual life cycles
III. have undulipodia but lack meiotic sexual life cycles
IV. have undulipodia and meiotic sexual life cycles

SUBKINGDOM (DIVISION) AMITOCHONDRIA

Phylum Pr-1 **Archaeprotista (III)** motile with no mitochondria

SUBKINGDOM (DIVISION) AMOEBAMORPHA

Phylum Pr-2 **Rhizopoda (I)** Amostigote amoebae, cellular slime molds

Phylum Pr-3 **Granuloreticulosa** (Foraminifera and unshelled relatives) **(IV)** reticulomyxids, foraminifera, chlorarachnids

Phylum Pr-4 **Xenophyophora (I)** barite skeleton deep sea protists

SUBKINGDOM (DIVISION) ALVEOLATA

Phylum Pr-5 **Dinomastigota (IV)** dinoflagellates

Phylum Pr-6 **Ciliophora (IV)** ciliates

Phylum Pr-7 **Apicomplexa (IV)** apicomplexan animal symbiotrophs

SUBKINGDOM (DIVISION) HETEROKONTA (STRAMENOPILES)

Phylum Pr-8 **Bicosoecida (III)** small mastigotes, some form colonies

Phylum Pr-9 **Jakobida (III)** bactivorous mastigotes, some loricate attached to sediment

Phylum Pr-10 **Proteromonadida (III)** small mastigotes, intestinal in animals

Phylum Pr-11 **Kinetoplastida (III)** kinetoplastids, most symbiotrophic mastigotes

Phylum Pr-12 **Euglenida (III)** euglenids

Phylum Pr-13 **Hemimastigota (III)** Gondwanaland mastigotes

Phylum Pr-14 **Hyphochytriomycota (III)** hyphochytrid water molds

Phylum Pr-15 **Chrysomonada (IV)** chrysophytes, golden-yellow algae

Phylum Pr-16 **Xanthophyta (IV)** yellow-green algae

Phylum Pr-17 **Phaeophyta (IV)** brown algae

Phylum Pr-18 **Bacillariophyta (IV)** diatoms, silica tests

Phylum Pr-19 **Labyrinthulata (IV)** slime nets and thraustochytrids

Phylum Pr-20 **Plasmodiophora (IV)** plasmodiophoran plant symbiotrophs

Phylum Pr-21 **Oomycota (IV)** oomycete water molds (=egg molds)

SUBKINGDOM (DIVISION) ISOKONTA

Phylum Pr-22 **Amoebomastigota Heterolobosea (III)** amoebomastigotes, heterolobosea
Naegleria, Paratetramitus, Vahlkampfia, Willaertia

Phylum Pr-23 **Myxomycota (IV)** plasmodial slime molds
Cercomonas, Echinostelium, Physarum, Stemonitis

Phylum Pr-24 **Pseudociliata (III)** polyundulipodiated animal symbiotrophs
Opalina, Stephanopogon, Zelleriella

Phylum Pr-25 **Haptomonada (III)** haptophytes, coccolithophorids
Emiliania, Phaeocystis, Prymnesium

Phylum Pr-26 **Cryptomonada (III)** cryptomonads, cryptophytes
Chilomonas, Cryptomonas, Cyathomonas, Nephroselmis

Phylum Pr-27 **Eustigmatophyta (III)** eye-spot algae
Chlorobotrys, Eustigmatos, Vischeria

Phylum Pr-28 **Chlorophyta (IV; green algae, plant ancestors)**
Acetabularia, Chlamydomonas, Volvox

SUBKINGDOM (DIVISION) AKONTA

Phylum Pr-29 **Haplospora (II)** haplosporan animal symbiotrophs

Phylum Pr-30 **Paramyxa (I)** cell-inside-cell marine animal symbiotrophs

Phylum Pr-31 **Actinopoda (II)** ray animalcules; acantharia, heliozoa, radiolaria

Phylum Pr-32 **Gamophyta (II)** conjugating green algae

Phylum Pr-33 **Rhodophyta (II)** red algae

SUBKINGDOM (DIVISION) OPISTHOKONTA

Phylum Pr-34 **Blastocladiomycota (IV)** polyzoosporic water molds
Phylum Pr-35 **Chytridiomycota (IV)** undulipodiated water molds, fungal ancestors
Phylum Pr-36 **Choanomastigota (III)** collared protists, animal ancestors

KINGDOM ANIMALIA
SUBKINGDOM (DIVISION) PLACOZOA (NO NERVES OR
ANTERO-POSTERIOR ASYMMETRY)

Phylum A-1 **Placozoa** free-living marine dorsal-ventral ciliated minimal animals
Phylum A-2 **Myxospora** reduced marine fish symbiotrophs

SUBKINGDOM (DIVISION) PARAZOA (NERVE NETS)

Phylum A-3 **Porifera** sponges
Phylum A-4 **Coelenterates** sea anemones, corals, hydroid-medusas
Phylum A-5 **Ctenophora** comb jellies

SUBKINGDOM (DIVISION) EUMETAZOA (NERVOUS AND
MUSCULAR SYSTEMS)

Phylum A-6 **Gnathostomulida** gnathostome worms
Phylum A-7 **Platyhelminthes** flatworms, flukes
Phylum A-8 **Rhombozoa** tiny animal symbiotrophs
Phylum A-9 **Orthonectida** tiny animal symbiotrophs
Phylum A-10 **Nemertina** nemertine worms
Phylum A-11 **Nematoda** nematode worms
Phylum A-12 **Nematomorpha** nematomorph worms
Phylum A-13 **Acanthocephala** spiny-headed worms
Phylum A-14 **Rotifera** rotifers; wheel-animals
Phylum A-15 **Kinorhyncha** kinorhynch worms
Phylum A-16 **Priapulida** priapulid worms
Phylum A-17 **Gastrotricha** gastrotrich worms
Phylum A-18 **Loricifera** loriciferans
Phylum A-19 **Entoprocta** entoprocts
Phylum A-20 **Chelicerata** horseshoe crabs, arachnids (spiders, scorpions, ticks)
Phylum A-21 **Mandibulata** insects, crustaceans, myriapods
Phylum A-22 **Annelida** earthworms, polychaete worms
Phylum A-23 **Sipuncula** sipunculid worms
Phylum A-24 **Echiura** echiurids
Phylum A-25 **Pogonophora** beardworms
Phylum A-26 **Mollusca** clams, snails, squid
Phylum A-27 **Tardigrada** water bears

Phylum A-28 **Onychophora** onychophorans

Phylum A-29 **Bryozoa** bryozoans, ectoprocts, moss animals

Phylum A-30 **Brachiopoda** lamp shells

Phylum A-31 **Phoronida** phoronid worms

Phylum A-32 **Chaetognatha** chaetognath worms

Phylum A-33 **Hemichordata** acorn worms

Phylum A-34 **Echinodermata** starfish, sea urchins

Phylum A-35 **Urochordata** tunicates, ascidians

Phylum A-36 **Cephalochordata** lancelets

Phylum A-37 **Craniata** vertebrates with skulls

KINGDOM FUNGI

Phylum F-1 **Microspora** fish symbiotrophs

Phylum F-2 **Zygomycota** coenocytic yoke fungi; molds

Phylum F-3 **Glomeromycota** arbuscular mycorrhizal (AM) root symbiotrophs

Phylum F-4 **Ascomycota** bladder fungi; yeasts, morels

Phylum F-5 **Basidiomycota** club fungi; mushrooms, puffballs

Phylum F-6 **Lichenes** fungal plus photosymbiotrophs

KINGDOM PLANTAE

SUBKINGDOM BRYATA

Phylum Pl-1 **Bryophyta** true mosses

Phylum Pl-2 **Hepatophyta** liverworts

Phylum Pl-3 **Anthocerophyta** hornworts

SUBKINGDOM TRACHEATA

Phylum Pl-4 **Lycophyta** club mosses

Phylum Pl-5 **Psilophyta** psilophytes

Phylum Pl-6 **Sphenophyta** horsetails

Phylum Pl-7 **Filicinophyta** ferns

Phylum Pl-8 **Cycadophyta** cycads; sago palms

Phylum Pl-9 **Ginkgophyta** maidenhair tree

Phylum Pl-10 **Coniferophyta** conifers

Phylum Pl-11 **Gnetophyta** gnetophytes

Phylum Pl-12 **Anthophyta** flowering plants

Glossary

Aleurone, the single layer of large cells under the bran coat and outside the endosperm of plant seeds; specific proteins and pigments are associated with this thin tissue.

Algae, photoautotrophic protoctists; all oxygenic phototrophs exclusive of cyanobacteria and plants; ecological term for aquatic oxygenic phototrophs.

Animalia, Kingdom, *see* Animals.

Animals, multicellular diploid organisms that develop from zygotes into embryos called blastulae. The embryo is formed by karyogamy after fertilization of egg cells by sperm cells. Animals develop by fusion of anisogametes, haploid gametes that differ in size. Females deliver mitochondria to the zygote in cytogamy. Meiosis produces gametes. Most are ingestive heterotrophs; some are absorptive heterotrophs. An estimated 10–30 million extant species in nearly forty major groups (phyla) in kingdom Animalia.

Archaea, Archaebacteria (the preferred term). One of two subkingdoms (domains) of kingdom Prokaryotae. Distinguished primarily by nucleotide sequences in ribosomal RNA.

Bacteria, microorganisms with prokaryotic cell organization; *see* Prokaryotes.

Bacteriocyte, also called mycetocyte. A specialized adipose animal tissue cell that contains endosymbiotic bacteria. Common in insects, for example, cockroaches, aphids, beetles.

Bacteriome, a specialized organ composed of bacteriocytes q.v. that contain endosymbiotic bacteria, usually in animals (mostly insects); also called mycetome.

Bioblast, an obsolete term for a prokaryotic cell or its part; in Kozo-Polyansky's book, this term is usually used for prokaryotic (non-nucleated) cells; *see* Ch. 1, n. 1.

Blepharoplast, kinetosome or group of developing kinetosomes; a conspicuous microtubule organizing center (MTOC) involved in cell division as determined by light microscopic observations of live cells (e.g., mastigotes, cycads, and ferns).

Blue-greens (blue-green algae; *sing.* blue-green alga), cyanobacteria, cyanophytes, Cyanophyceae. The old terms such as blue-green algae, cyanophytes, Cyanophyceae, have been replaced by "cyanobacteria" q.v., which recognizes the fundamental bacterial (prokaryotic) nature of these organisms.

Cell, a basic membrane-bounded unit of all living organisms in both prokaryotes (ancestral, most single-celled) and eukaryotes (organisms with nucleated cells, which originated from prokaryotes through endosymbiosis). From the nineteenth through the middle of the twentieth century, that is, for Kozo-Polyansky and his predecessors and contemporaries, the term "cell" was limited to nucleated (eukaryotic) cells.

Cell center, *see* Centrosome.

Centriole, barrel-shaped cell organelle, 0.25 µm (diameter) x 4 µm (length). Kinetosome lacking an axoneme; a [9(3)+0] microtubular structure that forms at each pole of the mitotic spindle during division in most animal cells. Observed to reproduce by a developmental cycle (for example, in which new centriole appears at right angles to the existing one).

Centrosome, an organelle that serves as the main microtubule organizing center (MTOC) in an animal cell as well as a regulator of cell cycle physiology; includes two centrioles.

Chimera, an organism merging two or more genetic types, formed by symbiosis, abnormal chromosome segregation, or artificial grafting; in Kozo-Polyansky's book, refers to a result of grafting in plants and transplants in animals.

Chlorobacteria, *see* Green bacteria.

Chlorophyll, green lipid-soluble pigments required for photosynthesis; all are composed of closed tetrapyrroles (porphyrins or chlorins) chelated around a central magnesium atom; comprise part of thylakoid membrane in all photosynthetic plastids.

Chlorophyll organelle, *see* Chloroplast.

Chloroplast, green plastid; membrane-bounded cell organelle containing lamellae (thylakoid membranes), chlorophylls *a* and *b*, usually carotenoids and other pigments, proteins, and nucleic acids in a nucleoid and ribosomes.

Chondriokont, an outdated term for mitochondria.

Chondriosome, an outdated term for mitochondria.

Chromatoplasm, peripheral colored protoplasm that contains chlorophyll, accessory pigments, and stored materials, for example, in cyanobacteria.

Chromidia, propagules, diameters from 2 to 5 μm, dark spherical bodies that originate from the nucleus, composed at least in part of chromatin; they pass to the cytoplasm and are seen released from cysts or amoebae, for example, in the vahlkampfid amoebomastigote *Paratetramitus jugosus.*

Chromiole, an obsolete term for small chromosome.

Chromosome, intranuclear organelle composed of chromatin (DNA, histone, and nonhistone protein) that contains most of the eukaryotic cell's genetic material; usually visible during the stages of mitotic nuclear division.

Consortium (*pl.* consortia), a group of individuals of different taxa (species to phyla) that live in close physical association. A physical association between the cells of two or more different types of microorganism that is stable and predictable. Often a syntrophy in microbiology, for example, *Chlorochromatium aggregatum.*

Cyanea, an outdated term for cyanobacteria.

Cyanobacteria, chlorophyll *a*, phycobiliprotein-containing, oxygenic photosynthetic bacteria; formerly called blue-green algae; phototrophic prokaryotes that use water (some may use sulfide) as an electron donor in the reduction of CO_2, produce oxygen in the light, contain thylakoids; they may be unicellular, multicellular (thallous), or filamentous. Some filamentous cyanobacteria differentiate nonphotosynthetic specialized cells (heterocysts) for nitrogen fixation; some have gliding motility. The most widespread phylum of phototrophic aerobic prokaryotes, cyanobacteria initiated the rise of gaseous oxygen in Earth's atmosphere at least two billion years ago.

Cyanophyceae, an outdated term for cyanobacteria.

Cytodes, an outdated term for prokaryotes. Swimming cytodes are referred to as "flagellated cytodes" by Kozo-Polyansky.

Cytoplasm, fluid portion of a cell that contains enzymes and metabolites in solution, and solid bodies (organelles) outside the nucleus and its nucleoplasm.

Endophyte, ecological term referring to the topology of symbiotic associates of plants; fungi, protoctists, or bacteria living within the tissue of plants or other photosynthetic organisms. Since "-phyte" means "plant" but traditionally has referred to fungi, protoctists, and bacteria, none of which are plants, the preferred term is endobiont (when the identity is in question), endosymbiotic bacteria, or other specific name. Commonly, "endophyte" refers to zygomycotous or ascomycotous fungi in plant tissue, especially of grasses.

Endosymbiont, endobiont. Ecological term referring to the topology of association of partners, a member of one species living inside the body or within a cell of a member of a different species. May be intracellular or extracellular.

Endosymbiosis, ecological term referring to the topology of an association of partners; the condition of one organism living inside another. Includes intracellular symbiosis (endocytobiosis) and inter-(extra-)cellular symbiosis.

Ergastoplasm, an obsolete term for endoplasmic reticulum (ER).

Eubacteria, one of two subkingdoms (domains) of kingdom Prokaryotae.

Eukarya, Superkingdom, *see* Eukaryotes.

Eukaryotes, all nucleated organisms, that is, those composed of nucleated cells associated with cytoskeletal systems. Nucleated (eukaryotic) cells evolved from integrated bacterial symbiosis; have chromosomal (rather than chromonemal) genetic organization. They contain microtubules as units of their cytoskeletal systems, including mitotic microtubules. Most have membrane-bounded hereditary organelles (such as mitochondria and plastids). Intracellular, microfilament- and microtubule-based motility (actin, myosin, tubulin-dynein/kinesin). Microtubule organizing centers (MTOC). Whole-cell (cytogamy) and nuclear fusion (karyogamy). Flexible steroid-containing membranes. Meiosis and fertilization cycles underlie Mendelian genetic systems. Levels of ploidy vary. Members of superkingdom Eukarya (kingdom Protoctista, kingdom Animalia, kingdom Fungi, kingdom Plantae).

Explantation, an outdated term that refers to the cultivation of cells, tissues, or organelles outside of the organisms from which the living material was taken.

Flagellate, obsolete term for mastigote; eukaryotic microorganism motile via undulipodia.

Flagellum (*pl.* flagella), bacterial flagellum; prokaryotic extracellular structure composed of homogeneous protein polymers, members of a class of proteins called flagellins; moves by a rotation at the base; relatively rigid rod driven by a rotary motor embedded in the cell membrane that is intrinsically nonmotile and sometimes sheathed. Undulipodium, so-called eukaryotic flagellum, by contrast, is an intrinsically motile intracellular structure used for locomotion and feeding in eukaryotes; composed of a standard arrangement of nine doublet microtubules and two central microtubules composed of tubulin, dynein, and approximately two hundred other proteins, none of them flagellin; no flagellum (but every undulipodium) is underlain by a kinetosome.

Fungi (*sing.* fungus), organisms with absorptive chemo-organoheterotrophic nutrition that are primarily terrestrial osmotrophs. Their chitinous-walled eukaryotic cells develop from uni- or multinucleated spores. Fungi lack both embryos and undulipodia at all stages of their life history. Fungi tend to be haploid cells or hyphae (threads) that sexually fuse to form dikarya that after karyogamy undergo zygotic meiosis to form resistant propagules (spores). An estimated 100,000 extant species of kingdom Fungi include molds, mushrooms, and yeasts.

Fungi, Kingdom, see Fungi.

Golgi apparatus (Golgi body, dictyosome), portion of the endomembrane system of nearly all eukaryotic cells visible with the electron microscope as a membranous structure of flattened saccules, vesicles, or cisternae, often stacked in parallel arrays; involved in elaboration, storage, and secretion of products of cell synthesis; prominent in many protoctists (e.g., in parabasalians the Golgi body is called the parabasal body due to its proximity to the kinetosome).

Grafting, a method of asexual plant propagation in agriculture and horticulture, where tissues of one plant are stimulated to fuse with those of another; one plant (stock or rootstock) is selected for its roots; the other plant (scion) is selected for its stems, leaves, flowers, or fruits.

Green animals, symbioses of animals (e.g., molluscs, platyhelminthes, coelenterates) with photobionts that confer on them the ability to photosynthesize (such as photosynthetic organelles, cyanobacteria or protoctists); for example, *Elysia* or *Symsagittifera (=Convoluta) roscoffensis.*

Green bacteria, one of two phyla of anoxygenic photosynthetic eubacteria (e.g., *Chlorobium*) that use hydrogen sulfide as hydrogen donor in photosynthesis.

Hypha (*pl.* hyphae), long, slender threadlike cells, walled syncytia, or parts of cells comprising the body of most fungi and many protoctists.

Kinetosome (also called by the outmoded term "basal body"), intracellular cytoplasmic organelle, not membrane-bounded, characteristic of the basal portion of the undulipodium in mastigotes, ciliated epithelium, and all other undulipodiated cells. They are cylindrical structures 0.25 m in diameter and up to 4 m long with microtubules organized in a standard [9(3)+0] array. All undulipodia are underlain by kinetosomes. Kinetosomes are required for and underlie the formation of undulipodia; kinetosomes differ from centrioles (which share cross-section characteristics of a circle of nine triplets of microtubules) in that from them extends the [9(2)+2] microtubular shaft.

Lichens, a phylum (Lichenes) of kingdom Fungi, comprised entirely of symbiotic organisms; most are symbioses of fungi with photosynthetic protoctists (algae) or cyanobacteria; in Kozo-Polyansky's book, lichens are grouped with plants ("lower plants").

Mastigote, a protoctist with undulipodia q.v., the preferred term for organisms still called "flagellates."

Metachromatic granules, granular cell inclusions in bacteria, protists, and tissues that, when stained, assume a color different from that of the dye. Known in some cases to be phosphate-rich storage organelles.

Microzymas, pleiomorphic organisms of A. Bechamps.

Mitochondrion (*pl.* mitochondria), membrane-bounded intracellular organelles containing enzymes and electron transport chains for oxidative respiration of organic acids and the concomitant production of ATP. Mitochondria contain DNA, messenger RNA, and small ribosomes and are thus capable of protein synthesis; mitochondria are nearly universally distributed in protoctists, fungi, plants, and animals but are notably absent in parabasalians, rhizopods, and other protoctist taxa.

Microtubule-organizing center (MTOC). Cell structure visible by electron microscopy. Includes kinetosome-centrioles, nuclear polar bodies, centrosomes, and other cell structures from which 24nm-diameter proteinaceous microtubules emerge.

Monera, *see* Prokaryotes.

Mucilage, mucous material that is generally composed of polysaccharides.

Mycelium (*pl.* mycelia), threadlike material (hyphae) that together forms a matted tissuelike structure that makes up the body of most fungi and some protoctists (e.g., chytridiomycotes, oomycotes).

Mycetocyte, see Bacteriocyte.

Mycetome, an organ formed by specialized cells (mycetocytes, bacteriocytes) that contain endosymbiotic bacteria (sometimes fungi), usually in animals (e.g., cockroaches, ants, beetles, lice, leeches, squid, etc.); see also bacteriome.

Mycorrhiza (*pl.* mycorrhizae), symbiotic associations (intracellular or extracellular) between fungi and plant roots. This fungal-root tissue forms organs crucial to mineral nutrient uptake in plant nutrition (e.g., phosphate and nitrate).

Nucleus, a membrane-bounded, spherical, DNA-containing organelle, universal in eukaryotes. Chromatin (DNA-protein) organized into chromosomes; site of DNA synthesis and RNA transcription. Nuclear membranes bear pores. Definitional for eukaryotes.

Organelle, any of a number of distinctive intracellular structures detected by microscopy in living (light microscopy) or fixed and stained (light and electron microscopy) cells. Some, such as mitochondria, nuclei, and plastids, are double-membrane-bounded and capable of division. Others, such as carboxysomes, ribosomes, and liposomes, are visualizable as locally high concentrations of specific enzymes or other molecules.

Organoid, an outdated term for organelle.

Parasitism, ecological association between members of different species in which one partner (usually the small form) is obligately or facultatively symbiotrophic and tends toward necrotrophy.

Physode, fucosan vesicle. Small, colorless vesicle in cells of phaeophytes (brown algae, often large seaweeds) that contain pigments and other components (e.g., fucosan, tannins, terpenes).

Plantae, Kingdom, *see* Plants.

Plants, photolithoautotrophic multicellular sexual organisms that develop from maternally retained diploid embryos. Nearly all are oxygenic photoautotrophs. Plant embryos form after fusion (fertilization) of mitotically produced gamete nuclei and grow into mature diploid sporophytes. On maturity these undergo meiosis and form haploid spores. Sporogenic meiosis produces male organisms or organs (antheridium; sperm-producing haploid plants) or female (archegonium; egg-producing haploid plant organs or organisms). Gametes formed in antheridia (male) and archegonia (female) are fertilized in archegonium. Alternating generations of haploid and diploid organisms. An estimated 600,000 extant species of plants are assigned to approximately twelve major groups (phyla) in kingdom Plantae.

Plastid, generic term for photosynthetic organelle in plants and protoctists (all algae). Bounded by double membranes, plastids contain the enzymes and pigments for photosynthesis, ribosomes, nucleoids, and other structures.

Prokarya, Superkingdom, *see* Prokaryotes.

Prokaryotae, Kingdom, *see* Prokaryotes.

Prokaryotes (Prokaryotae, Bacteria, Monera), bacteria; members of kingdom Prokaryotae (the only group of superkingdom Prokarya); single cells and multicellular organisms composed of cells with prokaryotic organization, which contain ultrastructurally visible nucleoids (genophore levels of organization) but lack membrane-bounded nuclei. Unidirectional cell-to-cell transfer of genophores (conjugation), or bacterial viruses, plasmids, or transforming principle DNA from solution. Cell walls of peptidoglycan, proteinaceous, or absent (but not cellulosic or chitinous). Ether- or ester-linked membrane lipids, without steroids, cytoskeletal system including cytoplasmic fusion and mitotic cell division absent. Flagellar rotary motor motility. Concept of ploidy inapplicable. From photolithoautotrophy to chemoorganoheterotrophy, all major modes of metabolism present in the group.

Protista (protists), microscopic protoctists q.v., obsolete formal kingdom name.

Protists, pleiomorphic live but healthy microorganisms suggested by Günther Enderlein to be present in normal vertebrate blood and tissues. Concept not adequately documented or accepted.

Protoctista, Kingdom, *see* Protoctists.

Protoctists (Protoctista), eukaryotic microorganisms (the single-celled protists) and their multicellular descendants. All eukaryotic organisms with the exception of animals (developing from diploid blastulae), plants (developing from embryos supported by maternal tissue), and fungi (developing from zygo-, asco-, or basidiospores) are protoctists. In the two-kingdom system, protoctists include "protozoans" and all "fungi" with mastigote stages as well as all algae (including kelps), slime nets, slime molds, and other obscure eukaryotes. Mitotic organisms capable of internal cell motility (that is, cyclosis, phagocytosis, pinocytosis). Many motile by undulipodia. Binary or multiple fusion. Meiosis and fertilization cycles absent or details unique to phylum. Photoautotrophs, ingestive and absorptive heterotrophs. Some 250,000 extant species are estimated, assigned to nearly forty large inclusive groups (phyla) in kingdom Protoctista. The smaller (microscopic) members of the kingdom Protoctista are often called "protists" or "protista."

Protoplasm, an outdated term for living fluid in cells; *see* Cytoplasm.

Protozoa (protozoans), an outdated formal term referring, in the two-kingdom classification, to a phylum in the animal kingdom consisting of large numbers of primarily heterotrophic, microscopic eukaryotes. Traditionally, include the smaller heterotrophic protoctists and their immediate photosynthetic relatives (e.g., phytomastigotes).

Pseudovitellus ("false yolk"), symbioorgan (mycetome) in aphids, replete with proteobacterial endosymbionts of the genus *Buchnera*.

Sponge-algae, animal-protoctist symbioses of sponges (Parazoa: Porifera) with oxygenic photobionts (e.g., green, golden yellow, or brown algae or cyanobacteria); abundant and varied components of marine, especially tropical, ecosystems.

Symbiogenesis, evolutionary origin of new behaviors, cells, organs, organisms, species, genera or other higher (more inclusive) taxa via symbiotic association between "differently named" organisms in which the selective advantage of the association can be identified.

Symbiont, members of a symbiosis, that is, organisms that have an intimate and protracted association with one or more organisms of a different species.

Symbioorgan, organ formed by symbiontic associates (e.g., mycetome).

Symbiosis, prolonged physical ecological association between two or more organisms of different kinds (e.g., in eukaryotes that belong to different species or other higher taxa). Levels of association in symbiosis may be behavioral, metabolic, gene product, or genomic; see also Endosymbiosis.

Undulipodium (*pl.* undulipodia), cilium, sperm tail; cell-membrane-covered motility organelles sometimes showing feeding or sensory functions; composed of [9(2)+2] microtubular axoneme usually covered by plasma membrane; limited to eukaryotic cells. Includes cilia and eukaryotic "flagella." Each undulipodium invariably develops from its kinetosome. Contrasts in every way with the prokaryotic motility organelle or flagellum, a rigid structure composed of a single protein (which belongs to the class of proteins called flagellins). Undulipodia in the cell biological literature are often referred to by the outmoded term "flagella" or "euflagella."

Vitules, fantastic elementary units of life suggested by A. Meyer (1921). Term not in use.

Zoochlorella (*pl.* zoochlorellae), green photosynthetic symbionts found in protoctists and animals. Although many belong to the genus *Chlorella* (e.g., unicellular chlorophyte algae in *Coleps hirtus, Hydra viridis,* and *Paramecium bursaria*), others belong to Prasinophyta, Dinomastigota, or other taxa; when only the green color is detected and the symbionts are unidentified to genus, the term "zoochlorella" is employed. When the unidentified photobiont is yellow or brownish, the term "zooxanthella" is used instead.

Zoogloea, a colony or mass of bacterial cells embedded in a viscous gelatinous substance of polysaccharide nature; also the name of a bacterial genus *Zoogloea* (fam. Rhodocycladaceae), where cells are embedded in such a matrix.

Zooxanthella (*pl.* zooxanthellae), yellowish or yellow-brown photosynthetic symbiont found in protoctists and animals. Although many belong to the dinomastigote group *Symbiodintium* (*Gymnodinium*), others belong to diatoms or other taxa; often the symbionts are unidentified to genus. See also Zoochlorella.

Index